T0321978

Revolutionizing Industrial Automation Through the Convergence of Artificial Intelligence and the Internet of Things

Divya Upadhyay Mishra
ABES Engineering College, Ghaziabad, India

Shanu Sharma
ABES Engineering College, Ghaziabad, India

A volume in the Advances in
Computational Intelligence and
Robotics (ACIR) Book Series

Published in the United States of America by
    IGI Global
    Engineering Science Reference (an imprint of IGI Global)
    701 E. Chocolate Avenue
    Hershey PA, USA 17033
    Tel: 717-533-8845
    Fax: 717-533-8661
    E-mail: cust@igi-global.com
    Web site: http://www.igi-global.com

Library of Congress Cataloging-in-Publication Data

Names: Mishra, Divya, editor. | Sharma, Shanu, DATE- editor.
Title: Revolutionizing industrial automation through the convergence of
  artificial intelligence and the internet of things / Divya Mishra and
  Shanu Sharma, editors.
Description: Hershey, PA : Engineering Science Reference, an imprint of IGI
  Global, [2022] | Includes bibliographical references and index. |
  Summary: "This book's aim is to include recent advancements in exploring
  and developing Artificial Intelligence of Things -powered strategies and
  mechanisms for future industrial automation and transforming industrial
  functions and architectures to help and improve various industrial
  operations"-- Provided by publisher.
Identifiers: LCCN 2022016707 (print) | LCCN 2022016708 (ebook) | ISBN
  9781668449912 (h/c) | ISBN 9781668449929 (s/c) | ISBN 9781668449936
  (ebook)
Subjects: LCSH: Automation. | Artificial intelligence. | Internet of
  things.
Classification: LCC T59.5 .R49 2022 (print) | LCC T59.5 (ebook) | DDC
  006.3--dc23/eng/20220624
LC record available at https://lccn.loc.gov/2022016707
LC ebook record available at https://lccn.loc.gov/2022016708

This book is published in the IGI Global book series Advances in Computational Intelligence and Robotics (ACIR) (ISSN: 2327-0411; eISSN: 2327-042X)

Advances in Computational Intelligence and Robotics (ACIR) Book Series

ISSN:2327-0411
EISSN:2327-042X

Editor-in-Chief: Ivan Giannoccaro University of Salento, Italy

MISSION

While intelligence is traditionally a term applied to humans and human cognition, technology has progressed in such a way to allow for the development of intelligent systems able to simulate many human traits. With this new era of simulated and artificial intelligence, much research is needed in order to continue to advance the field and also to evaluate the ethical and societal concerns of the existence of artificial life and machine learning.

The **Advances in Computational Intelligence and Robotics (ACIR) Book Series** encourages scholarly discourse on all topics pertaining to evolutionary computing, artificial life, computational intelligence, machine learning, and robotics. ACIR presents the latest research being conducted on diverse topics in intelligence technologies with the goal of advancing knowledge and applications in this rapidly evolving field.

COVERAGE

- Artificial Intelligence
- Heuristics
- Evolutionary Computing
- Intelligent Control
- Artificial Life
- Fuzzy Systems
- Computer Vision
- Brain Simulation
- Cyborgs
- Neural Networks

IGI Global is currently accepting manuscripts for publication within this series. To submit a proposal for a volume in this series, please contact our Acquisition Editors at Acquisitions@igi-global.com or visit: http://www.igi-global.com/publish/.

Titles in this Series

For a list of additional titles in this series, please visit:
www.igi-global.com/book-series/advances-computational-intelligence-robotics/73674

Convergence of Big Data Technologies and Computational Intelligent Techniques
Govind P. Gupta (National Institute of Technology, Raipur, India)
Engineering Science Reference • © 2023 • 335pp • H/C (ISBN: 9781668452646) • US $270.00

Design and Control Advances in Robotics
Mohamed Arezk Mellal (M'Hamed Bougara University, Algeria)
Engineering Science Reference • © 2023 • 320pp • H/C (ISBN: 9781668453810) • US $305.00

Handbook of Research on Applied Artificial Intelligence and Robotics for Government Processes
David Valle-Cruz (Universidad Autónoma del Estado de México, Mexico) Nely Plata-Cesar (Universidad Autónoma del Estado de México, Mexico) and Jacobo Leonardo González-Ruíz (Universidad Autónoma del Estado de México, Mexico)
Information Science Reference • © 2023 • 335pp • H/C (ISBN: 9781668456248) • US $315.00

Applying AI-Based IoT Systems to Simulation-Based Information Retrieval
Bhatia Madhulika (Amity University, India) Bhatia Surabhi (King Faisal University, Saudi Arabia) Poonam Tanwar (Manav Rachna International Institute of Research and Studies, India) and Kuljeet Kaur (Université du Québec, Canada)
Engineering Science Reference • © 2023 • 300pp • H/C (ISBN: 9781668452554) • US $270.00

Computer Vision and Image Processing in the Deep Learning Era
A. Srinivasan (SASTRA Deemed To Be University, India)
Engineering Science Reference • © 2022 • 325pp • H/C (ISBN: 9781799888925) • US $270.00

For an entire list of titles in this series, please visit:
www.igi-global.com/book-series/advances-computational-intelligence-robotics/73674

701 East Chocolate Avenue, Hershey, PA 17033, USA
Tel: 717-533-8845 x100 • Fax: 717-533-8661
E-Mail: cust@igi-global.com • www.igi-global.com

Table of Contents

Detailed Table of Contents

Chapter 1

Sandhya Avasthi, ABES Engineering College, India
Tanushree Sanwal, Krishna Institute of Engineering and Technology,
   India
Shivani Sharma, Indraprastha Engineering College, India
Shweta Roy, ABES Engineering College, India

The vehicular ad-hoc network (VANET) has emerged as the most sought-after technology due to its wide range of applications. In this modern era, people are looking for intelligent systems that include intelligent transport systems, which is not possible without the use of modern techniques such as the internet of things (IoT), VANETs, and cloud computing. With the growing demand for luxury cars, and people's need for safety, the ad-hoc car network is experiencing growth. The connectivity between the vehicle is possible using sensor communication among vehicles in the network. For transmitting forthcoming traffic information in case of a vehicle accident, car connectivity plays a vital role to connect to other vehicles so that appropriate actions can be taken. In this chapter, applications, protocols, trust models, and challenges of VANETS are discussed. The purpose of this chapter is to discuss and elaborate on various aspects related to VANETs and IoT. In addition, the chapter discusses characteristics, challenges, and security concerns in VANETs-based applications.

Chapter 2

Subha Karumban, Alagappa University, India

Shouvik Sanyal, Dhofar University, Oman

Madan Mohan Laddunuri, Malla Reddy University, India

Vijayan Dhanasingh Sivalinga, Arupadai Veedu Institute of Technology, India

Vidhya Shanmugam, REVA University, India

Vijay Bose, Vaagdevi College of Engineering, India

Mahesh B. N., Reva University, Bengaluru, India

Ramakrishna Narasimhaiah, School of Humanities and Social Sciences, Jain Deemed to be University, Bengaluru, India

Dhanabalan Thangam, Acharya Institute of Graduate Studies, Bengaluru, India

Satheesh Pandian Murugan, Arumugam Pillai Seethai Ammal College, India

Industries in recent days has been facing several issues including a dearth of labor, moribund joblessness rates, and towering labor turnover. This is projected to increase further in the upcoming days with the rise of the aged population. All these challenges can be addressed by the technology called industry automation. But there is an acuity that industrial automation will lead to job losses. However, in ingenious automation technology, lots of positive outputs can be achieved such as higher productivity and enhancement in yield. The role of automation in health and safety is awesome, but the brunt of this technology on jobs and efficiency worldwide has not been studied fully. In addition, automation technology safeguards the workers from highly dangerous work zones such as mines, space research, and underwater research. Thus, industrial automation is ready to serve humankind and business in various ways. This has been explained in this chapter.

Chapter 3

Aditya Saxena, Amity University, Noida, India

Devansh Chauhan, Amity University, Noida, India

Shilpi Sharma, Amity University, Noida, India

This chapter discusses the various impacts of industrial automation and artificial intelligence in supply chains with the onset of COVID-19. The term industrial automation is influenced by rapid globalization and the various industrial revolutions that have caused the dire need for automation of industrial tasks to reduce human efforts. The chapter dives into the multiple fields affected by COVID-19 and how automation was used to deal with the situation, stabilize the supply chains, and maintain the profitability of organizations. Digital globalisation has led to the development of global supply chains. The use of technologies and cognitive automation and its effects have been discussed in the chapter. Machine learning has been used to get insight into the factors that affect supply chains and help their functioning.

Eduardo José Villegas-Jaramillo, Universidad Nacional de Colombia -
Sede Manizales, Colombia
Mauricio Orozco-Alzate, Universidad Nacional de Colombia - Sede
Manizales, Colombia

Convolutional neural networks and their variants have revolutionized the field of image processing, allowing to find solutions to various types of problems in automatic visual inspection, such as, for instance, the detection and classification of surface defects in different types of industrial applications. In this chapter, a comparative study of different deep learning models aimed at solving the problem of classifying defects in images from a publicly available glass surface dataset is presented. Ten experiments were designed that allowed testing with several variants of the dataset, convolutional neural network architectures, residual learning-based networks, transfer learning, data augmentation, and (hyper)parameter tuning. The results show that the problem is difficult to solve due to both the nature of the defects and the ambiguity of the original class labels. All the experiments were analyzed in terms of different metrics for the sake of a better illustration and understanding of the compared alternatives.

Revathi A., VISTAS, India
Poonguzhali S., VISTAS, India

Agriculture is not all about food and includes production, promotion, filtering, and sales of agricultural products and provides better employment opportunities to many people. Nowadays, precision agriculture is gaining more popularity, and its main goal is the availability to all common people at low cost with maximum crop productivity. It also helps in protecting the environment. IoT, internet of things, a technology that is developing in modern society, can be applied to agriculture. At present, IoT-enabled technology in agriculture has developed to a greater extent, particularly a drastic development of unmanned aerial vehicles (UAVs) and wireless sensor networks (WSN), and could lead to valuable but cost-effective applications for precision agriculture (PA), including crop monitoring with drones and intelligent spraying tests. In this chapter, the authors explore the various applications of artificial intelligence of things (AIoT) and provide detailed explanation on how AIoT may be implemented in agriculture effectively. Moreover, they highlight crucial future research and directions for AIoT.

Chapter 6

Indu Malik, Gautam Buddha University, India
Anurag Singh Baghel, Gautam Buddha University, India

Presently there is a massive enhancement in technologies, and a lot of things, appliances, and techniques are accessible in the agriculture sector. One of the famous techniques is known as IoT. Several applications of IoT are evident in the field of agriculture for the benefit of the farmers and in turn for the successful development of the nation. IoT is used in agriculture to improve productivity, efficiency, and the global market. It also helps farmers to reduce manpower, cost, and time. In this chapter, the authors are discussing IoT with the cloud for enhancing smart farming. Smart farming is a concept that is focused on providing the agricultural industry with the infrastructure to leverage advanced technology including big data, the cloud, and the internet of things (IoT) for tracking, monitoring, automating, and analyzing operations. IoT with the cloud is used in various fields of agriculture to improve time efficiency, water management, crop monitoring, and land management. It protects the crops from pests and is also used to control pesticides in agriculture.

Chapter 7

Shanu Sharma, Department of Computer Science and Engineering,
ABES Engineering College, India
Tushar Chand Kapoor, Department of Computer Science and
Engineering, ASET, Amity University, Noida, India
Misha Kakkar, Department of Computer Science and Engineering,
ASET, Amity University, Noida, India
Rishi Kumar, Universiti Teknologi Petronas, Malaysia

In this chapter, an optimized parking framework, AUTOHAUS, is presented that focuses on three aspects (i.e., automation, security, and efficient management of parking space). A combination of advanced technologies is used to design the proposed framework. AUTOHAUS provides two ways of security implementation such as authorized QR codes and OTPs (one-time passwords). Furthermore, for efficient management of parking spaces, the still images of the front and side view of the car are used to extract the license plate and size of the car for effective allotment of parking space based on the size of the car. This proposed system can reduce human effort to a great extent and can also be used as a path-breaking technique in parking and storage management.

Chapter 8
Indu Malik, ABES Engineering College, India
Arpit Bhardwaj, BML Munjal University, India
Harshit Bhardwaj, Galgotias University, India
Aditi Sakalle, Gautam Buddha University, India

The smart home achieved commerciality in the previous decades because it has increased comfort and quality of human life. Using AI and IoT technology, humans get notifications about the unplaced (not having or assigned to a specific place) things or devices at home. In this chapter, the authors introduce the concept of a smart home with the integration of IoT (internet of things), cloud, and AI. A smart home can be controlled by smartphones, tablets, or PC. The smart home is a collection of smart devices; smart devices have sensors and actuators. All these devices are connected, and they are controlled by the central unit. The sensor is used to collect data. The smart home is a net of sensors, different types of sensors used to create a smart home. Particular sensors perform particular tasks like biometric sensors used for security. In this chapter, the authors discuss remote access of devices, increased computation power, and data storage. Home automation gives you access to an on and off home light, lock and unlock the door, and switch on and switch off the AC and TV.

Chapter 9
Divya Upadhyay, ABES Engineering College, Ghaziabad, India
Ayushi Agarwal, ABES Engineering College, Ghaziabad, India

The 21st century is the era of the digital world and advanced technologies. This chapter contributes to the Swachh Bharat mission by presenting the concept of smart bin using IoT. The Smart bin presented in this chapter is GPS-enabled and comprises sensors and a camera. A prototype for the proposed model is analysed, and network architecture is designed to communicate the critical information. The proposed system will update the status and condition of the bin to the nearest authority to improve the city's pollution and cleanliness. The prototype is deployed using a microcontroller Raspberry Pi and Google Maps to obtain the bins' real-time location. IoT fill-level sensors will help the garbage carrying truck in identify the nearest empty container without wasting time and resources. Google Maps will help in sensing the optimised routes to the drivers. The microcontroller will be used to integrate the different devices and cameras to provide real-time bin collection, overflowing/under flowing state and tracking information, and suggestions and notifications for effective disposal.

Chapter 10
An Unsupervised Traffic Modelling Framework in IoV Using Orchestration

Divya Lanka, Pondicherry Technological University, India
Selvaradjou Kandasamy, Pondicherry Technological University, India

Presenting better traffic management in urban scenarios has endured as a challenge to reduce the fatality rate. The model developed consolidates an evolving methodology to handle traffic on roads in an abstract way in internet of vehicles (IoV) orchestrated with unsupervised machine learning (USL) techniques. At first, the roads are sliced into segments with roadside units (RSU) that provide vehicle to everything (V2X) communication in a multi-hop manner and examine traffic in real-time. The unique nature of the proposed framework is to introduce USL into the RSU to learn about traffic patterns. The RSU upon learning the traffic patterns applies the Gaussian mixture model (GMM) to observe the variation in the traffic pattern to immediately generate warning alerts and collision forward messages to the vehicles on road. The application of USL and GMM ensures speed control of vehicles with traffic alerts, thereby deteriorating the fatality rates.

Chapter 11
Exploring CNN for Driver Drowsiness Detection Towards Smart Vehicle

Pushpa Singh, GL Bajaj Institute of Technology and Management,
Greater Noida, India
Raghav Sharma, KIET Group of Institutions, India
Yash Tomar, KIET Group of Institutions, India
Vivek Kumar, KIET Group of Institutions, India
Narendra Singh, GL Bajaj Institute of Technology and Management,
India

Driver drowsiness is one of the major problems that every country is facing. The ICT sector is continuously investing in the automaker industry worldwide to bring about digital transformation in existing vehicles and driving. The smart behavior of vehicles is becoming possible with the convergence of intelligent manufacturing, AI, and IoT. In this chapter, the authors are presenting a framework for efficient detection of driver's drowsiness by utilizing the power of deep learning technology. The use of convolution neural network (CNN) is explored, and the system is developed and tested using different activation functions. The proposed driver drowsiness framework is able to signify the drowsiness state of the driver and to automatically alert the driver. The accuracy of the proposed model is compared at different activation functions such as ReLu, SeLu, Sigmoidal, Tanh, and SoftPlus, and higher accuracy is achieved with ReLu as 98.21%.

    Sachin Kumar Yadav, HMRITM, Delhi, India
    Devendra Kumar Misra, ABES Engineering College, India

A new era of technology has evolved over the years. Reliability has grown because of its accessibility, speed, and efficiency. This has enabled us to maintain the environment ranging from reducing pollution to low consumption of electricity. Nowadays we are relying so much on digital gadgets that it has started to play an important role in our daily lives, as these digital gadgets can connect with many other devices, making us perform our tasks more easily. With the help of the home automation system, we can not only connect several different devices, but we can also reduce the power consumption. The hallmark of home automation in this proposed work is a remote control, which is done either through a mobile application or a voice assistant. In this system, the mobile application is used for the Android system and for connectivity, Bluetooth has been used as the medium to control the devices. The mobile application allows them to control the devices in real-time. In addition, a capacitive switch that replaces the toggle switch has been used.

Foreword

With the development of innovative technologies, "Industrial Automation" is obtaining extraordinary attention from academia, government, researchers, and the various industrial communities. In today's era, the technical innovations in advanced technologies such as Deep learning, AIoT, Blockchain, etc. have unlocked the potential for bringing intelligent automation and efficiency to the industries to control their various operations. On one side AIoT is taking automation to a new level, and making smart systems truly smart, on the other hand, advanced learning methods such as deep learning are providing ways to handle the large amount of data populated through these IoT devices. These new technologies have the capability of providing an efficient and secure way to industries for managing and developing products intelligently.

Various industries such as health care, agriculture, automation, governance, etc. are absorbing new methods for reducing costs, increasing productivity along with safety for human workers as well as for the industrial appliances. The execution of technology innovation is vital for industrial automation and efficient management and progress in this direction is the need of the hour for sustainable development of society.

With *Revolutionizing Industrial Automation Through the Convergence of Artificial Intelligence and the Internet of Things*, the editors and contributors are providing a glimpse of the power of this convergence in different domains such as Manufacturing, Transportation, Smart Cities, Supply Chain Management, etc. The discussion on the usage of this convergence in these domains is the need of the era. It can help various researchers to further explore its applicability in undiscovered areas and to highlight major issues during its implementation in various sectors.

Priya Ranjan
Bhubaneshwar Institute of Technology, Odisha, India

Preface

In recent times, the Internet of Things (IoT) has converted an innovative world vision into reality with actual data and numerous services. On the other hand, Artificial Intelligence (AI) fits very well with IoT to enable high-level cognitive processes like perceiving, thinking, problem-solving, learning, and decision-making capabilities. From a business perspective also, IoT and AI are two increasingly dominant technology trends. Together, these two technologies – AI and IoT – promise to go beyond mere buzzwords and redefine the future of Industrial Automation.

The confluence of Artificial Intelligence and Internet of Things (AIoT) technologies has given rise to a new digital solution category in every domain. AIoT has a lot of scopes, from connected equipment in the factory to consumer devices to industrial assets like robots, machines, and workers in intelligent and smart factories. Bringing a revolutionary change to automation industries helps intensify operations and increase production. The execution of AIoT concepts and technology is vital for industrial automation. Its main aim is to create an intelligent and constant connection between manufacturing inputs and outputs. At the same time, the whole thing involves smart connectivity between complex and sophisticated devices. AIoT is built for industries looking for improved practices to connect their evolving workforce to data-driven decision tools, digitally augment work and business processes, and better use industrial data already collected.

This book presents recent advancements in exploring and developing AIoT-powered strategies and mechanisms for future automation and transforming industrial functions and architectures to help and improve various operations. This book focuses on automation in various industries such as agriculture, home automation, supply chain, defect detection, parking management, and manufacturing industries.

The seamless integration between AI software and IoT hardware will be the technology that revolutionizes the smart industries era due to the immediate need for machine intelligence and hyper-automation. This book presents an architectural challenge, characteristics & approaches for AIOT in Industrial automation in chapter 1. The purpose of this chapter is to examine and elaborate on various aspects related to VANETs and IoT. The chapter discusses characteristics, challenges, and security concerns in VANETs-based applications.

A classic example of applying AIoT to the industry is at a large biomass production facility in Vietnam. The system was originally designed and built with traditional PLC/HMI/SCADA technology and provided baseline functionality. However, the facility owners were looking for a way to improve operations without getting into complex and expensive engineering cycles. The role of automation in health and safety is awesome, but it is not studied the brunt of this technology fully on jobs and efficiency worldwide. Besides, automation technology safeguards the workers from highly dangerous work zones such as mines, space research, and underwater research. Thus, industrial automation can serve a lot of humankind and business in various ways, and the book covers the same acceptance of industrial automation in chapter 2.

AI and IoT will Revolutionize supply chain management and Industrial Automation Markets. AI in SCM solutions will reach $15.5B globally by 2026. The Asia Pac region is the largest and fastest growing for AI in supply chain management. Chapter 3 demonstrates the effect of Industrial Automation and Artificial Intelligence on the supply chain with the onset of covid. Another chapter 4 is included in catering for the problem faced by the glass industry. It presents convolutional neural networks and deep learning techniques for Glass Surface Defect Inspection.

Agriculture is the most important industry in India. Chapter 5 presents the role of AIoT in agriculture. Autonomous drones will allow farmers to capture images of crops and monitor their conditions remotely. Using the UAVs, growers can apply crop treatments like pesticides and fertilizers from the air. AI-powered cameras can get mounted on drones and deployed to large-scale farms. The cameras will help growers detect issues with crops, count fruits, and even forecast crop yield. AI also allows the automation of other farming activities, like harvesting, seeding, weeding, and crop sorting. Australia's farm leverages AI and robots' power to conduct hands-free farming. Farmers can use powerful and efficient AI platforms to manage all the data. The power of data can help reduce the cost of labour, increase yield production, and reduce the environmental footprint of farming.

Furthermore, it can help farmers evaluate their farming strategies and resource management for maximum productivity and profits. This book explores the various applications of Artificial Intelligence of Things (AIoT) and explains how AIoT may be implemented in agriculture effectively. Moreover, chapter 5 highlights crucial future research and direction for AIoT in agriculture. Chapter 6 presents a study on IoT in Agriculture to Implement Smart Farming.

The problem of parking in urban cities is a big issue these days. These industries are working on designing parking management tools to automate these problems.

This book presents an approach to using AIOT to solve the parking problem through chapter 7.

Smart homes have achieved commerciality in the previous decades because it has increased comfort and quality of human life. Chapter 8 presents remote access to devices for home automation. As stated by Cisco: by 2050, approx. Fifty million users and devices are connected using the internet. Chapter 9 presents an approach to the "Swachh Bharat Mission" by presenting the concept of Smart Bin using IoT.

AIoT is efficiently deployed to empower the automobile industries and road management induteries. Chapter 10 discussed an unsupervised traffic modelling framework in IoV using orchestration of road slicing. Driver Drowsiness is one of the major issues in the smart vehicle industry. In Chapter 11, the authors presented a framework for efficient detection of driver's drowsiness by utilizing the power of deep learning technology. Chapter 12, presents an IoT and voice-based framework for a secured and automated workplace environment.

Most importantly, AIoT can analyze the vast amount of telemetry data from connected devices in real-time. Unlike traditional IoT systems, which are reactive and designed to respond to urgent events by sending alerts to stakeholders, AIoT architectures are proactive, capable of predicting equipment failures and shutting down faulty machines before the event occurs. Manufacturers must upgrade their network infrastructure to utilize intelligent and connected systems fully.

This book has many use cases for AIoT in different industries, which totally depend on a user's business objectives. This book can help its readers to improve their overall productivity. While there will be challenges along the way, the benefits of implementing a combination of IoT and AI will outweigh them.

Divya Upadhyay Mishra
ABES Engineering College, Ghaziabad, India

Shanu Sharma
ABES Engineering College, Ghaziabad, India

Acknowledgment

First of all, we would like to thank the publisher for trusting us and providing us with the opportunity to collaborate on this book. Your support and guidance during the complete book processing cycle were extraordinary.

Editors are grateful to the Editorial Board Members and Technical Review Committee Members of the book, who helped us at various steps of completion of the book. Their expert comments were valuable and contributed to the overall enhancement of the book chapters. We express sincere gratitude for your persistent support.

The completion of this book project could not have been accomplished without our sincere contributors. We would like to thank them to allow us to showcase their work and expertise through this book.

Last but not least, we want to express our gratitude to "ABES Engineering College, Ghaziabad," for supporting us and allowing us to work on this book project.

Thanks

Divya Upadhyay Mishra

Shanu Sharma

Chapter 1
VANETs and the Use of IoT:
Approaches, Applications, and Challenges

Sandhya Avasthi
https://orcid.org/0000-0003-3828-0813
ABES Engineering College, India

Shivani Sharma
Indraprastha Engineering College, India

Tanushree Sanwal
Krishna Institute of Engineering and Technology, India

Shweta Roy
ABES Engineering College, India

ABSTRACT

The vehicular ad-hoc network (VANET) has emerged as the most sought-after technology due to its wide range of applications. In this modern era, people are looking for intelligent systems that include intelligent transport systems, which is not possible without the use of modern techniques such as the internet of things (IoT), VANETs, and cloud computing. With the growing demand for luxury cars, and people's need for safety, the ad-hoc car network is experiencing growth. The connectivity between the vehicle is possible using sensor communication among vehicles in the network. For transmitting forthcoming traffic information in case of a vehicle accident, car connectivity plays a vital role to connect to other vehicles so that appropriate actions can be taken. In this chapter, applications, protocols, trust models, and challenges of VANETS are discussed. The purpose of this chapter is to discuss and elaborate on various aspects related to VANETs and IoT. In addition, the chapter discusses characteristics, challenges, and security concerns in VANETs-based applications.

DOI: 10.4018/978-1-6684-4991-2.ch001

INTRODUCTION TO VANETs AND IOT

Internet of Things (IoT), which is fast emerging as a powerful technology, coupled with intelligent and integrated sensor network systems and domestic sensor networks are anticipated to have an impact on people's daily lives and stimulate a significant market shortly. The Internet of Things is a network of connected things such as mobile devices, smart sensors in the vehicle, digital machines, other computing devices, and even people. IoT has expanded into the field of smart vehicles and turned into something called as Internet of Vehicles (IoV) (Ahmood, 2020). The proper functioning of IoV is based on Vehicular Ad-Hoc Network. An integral component of any "Intelligent Transport Systems" (ITS) is VANET, who's growth is accelerating fast. VANET and IoT are currently the most crucial components of the "Intelligent Transport System" (ITS). The research study in the past decade on VANET and IoT indicates that both have a significant impact on intelligent transportation systems. Road accidents, congestion, fuel consumption, and environmental pollution have all become major global challenges as the number of automobiles increases. In both developed and developing countries, traffic accidents frequently cause massive loss of property and human life. ITS formed and implemented VANETs to provide infrastructure for transportation of all types. The importance lies in dealing with prevalent issues in transportation to make the journey for everyone safer, effective, hassle-free, and enjoyable (Hossain et al.,2012).

In any Mobile ad-hoc network, an integral part is VANET and therefore, nodes operate inside the networking region and devices operating in that area. Transfer of information with one another through single-hop or multi-hop through a road-site unit (RSU) (Patel et al.,2015). VANET's advantage is to improve vehicle safety by switching caution messages among vehicles. VANET's main concern is to improve passenger safety and the exchange of security messages between locations. Security is very important for VANETs because of the scarcity of centralization, and powerful arrangement of nodes leading to extreme difficulty in recognizing nodes or network vehicles that are dangerous and, malicious (Hussain et al., 2015). Vehicles are in direct contact with some other vehicle, if in case, there exists the availability of wireless connection; it is called a single vehicle to vehicle (V2V). All the motor vehicles operating within the network are connected to Road-Side-Unit (RSU) which further expands the network vehicle communication by sending a message and getting details from them.

In VANETs, the two primary types of applications are *safety and non-safety applications*. For purpose of sending safety messages, safety applications are used such as warning messages. Warning messages help and assist vehicles on the road in case of collisions that saves a life. Messages regarding road safety include reports of car accidents, traffic jams, road construction, and alerts from emergency vehicles

(Hartenstein & Laberteaux,2010). Non-safety applications, on the other hand, make driving easier and more comfortable. Traffic management and information and entertainment are the two types of non-safety applications. Traffic management enhances smooth traffic flow and eliminates traffic congestion by incorporating advanced applications. The applications, known as infotainment applications entertain passengers by providing Internet access, the ability to store data, stream videos, make video calls, and information related to maps through GPSs (Global Position Systems). Unlike safety applications, these applications are not required to be extremely reliable and quick. In case of crucial events, accident reports can be disseminated fast and this is reliable over VANETs. Even though VANETs can be used to disseminate event information, it is still difficult to deliver critical messages to the appropriate location and time in a dynamic vehicular setting.

Vehicular Ad-hoc Networks

People in the modern world need cars more than anything else. Accidents are happening more often, even though safety devices like advanced braking systems, rear-view cameras, seatbelts, etc. are getting better and more people are using private transportation. Several studies have shown that about 60% of road accidents can be avoided if the driver gets a warning message a few seconds before the crash. Intelligent Transportation Systems (ITS) constitute the present-day world. The nodes in a VANET are highly dynamic, so the network topology changes rapidly. To provide intelligent driving, safe navigation, entertainment, and emergency applications, VANETs are being utilized in implementation. VANETs are network-dependent that makes them very different from Intelligent Transportation Systems (Shinde & Patil, 2010, Tanuja et al.,2015, Zeadally et al.,2012). Table 1 displays the various wireless access technologies that are compatible with VANET.

This chapter presents an overview of VANETS, protocols, trust model, IoT technologies, and their useful application in various aspects of life. Important aspects related to the architecture of VANETs, characteristics, protocols, and challenges have been discussed in this chapter. In addition, security and privacy issues along with future advancement is provided.

Intelligent Transportation Systems and Internet of Vehicle

Integration of information and communication technologies has proved efficient in traffic management systems, these are referred to as "Intelligent Transportation Systems" (ITS). Such a system aims to provide safety, sustainability of transportation networks, and efficiency. In addition to this, ITS helps in reducing traffic congestion and improving the experiences of drivers. In addition, they hope to reduce the

Table 1. Wireless technologies and comparison

SN	Wireless Access Techniques	Coverage	Speed	Density	Throughput	Latency	Reliability
1	C-V2X	High	High	High	High	Low	High
2	DSRC	Low-moderate	High	Low	Low	Moderate	no
3	4G-LTE	Higher	Low	High	Moderate	Moderate	High
4	WiFi	Moderate	Low	Low	Low	Moderate	No
5	UWB	Moderate	Low	Low	Low	Low	No
6	ZigBee	Low	Low	Low	Low	Low	no
7	BlueTooth	Low	Low	Low	Low	Low	no

amount of time that drivers spend stuck in traffic. The possibilities are endless and have no bounds.

Sensors, software, and the technologies that connect and exchange data among these components are all part of the "Internet of Vehicles" (IoV) (Lee,2016). IoV is an extension of vehicle-to-vehicle (V2V) communication that improves assistance systems for fully autonomous driving by raising the vehicles' artificial intelligence awareness of their surroundings (including other vehicles and driving-related smart devices). The following components make up the IoV: vehicles (as ad hoc network nodes), roadside units (RSU), infrastructure (such as road and traffic-related sensors, signals, and smart objects), personal electronics (such as smartphones and PDAs), and people (e.g., drivers) (Kumar, 2020).

The "Internet of Autonomous Vehicles" (IoAV) is expected to develop from Vehicular Ad Hoc Networks ("VANET," a type of mobile ad hoc network used for communication between vehicles and roadside systems). Autonomous, connected, electric, and shared (ACES) Future Mobility will be enabled in part by IoV. Real-time analytics (Avasthi et al.,2022), commodity sensors, and embedded systems all contribute to the development of road vehicles. For the IoV ecosystem to function as a whole, it requires modern architectures and infrastructures that distribute computational load across multiple network nodes (Khelifi,2019, Avasthi et al., 2021).

VANETS AND ITS ARCHITECTURE

An overview of the main architectural components of VANET is provided here from a domain view. Other aspects such as the interaction between components and communication architecture is explained too.

Main Components

As per IEEE 1471-2000 (Maier et al.,2001) and ISO/IEC 42010 (Maier et al., 2002) architecture standard guidelines, one can implement the VANETs system by various entities taking part in it. The mobile domain, the infrastructure domain and the generic domain are three groups into which entities are divided (Schroth et al., 2012). There are two mobile domains, the first is a vehicle and the second is a mobile device as given in Figure 1. Devices such as cars, buses, and other transport mediums generally become part of the vehicle domain. The devices like navigation devices and smartphones are part of the mobile device domain. Another category is the infrastructure domain which includes central infrastructure and roadside infrastructure. Elements like traffic signals, light posts, and roadside units are called roadside infrastructure whereas traffic management centers (TMC) and vehicle management centers are called central infrastructure (Baldessari et al.,2007).

The development of VANET architecture, however, differs per region. The CAR-2-CAR communication consortium is pursuing a standard design for the CAR-2-X communication system that is a little bit different. The primary force behind vehicular communication in Europe is the CAR-2-CAR communication consortium (C2C-CC), which published its "manifesto" in 2007. The in-vehicle, ad hoc, and infrastructure domains of this system design are separated. The in-vehicle domain consists of one or more application units and an on-board unit (OBU) (AUs). They often use wired connections, although they also employ wireless connections on occasion. The ad hoc domain, on the other hand, is made up of cars carrying OBUs and roadside units (RSUs) (Faezipour et al.,2012). A RSU is a static node, on the other hand OBU is a mobile node. RSU get the connectivity through internet via gateway, these units communicate with each other directly or through multihop. In the infrastructure domain, the access is done through RSUs and hot spots (HSs). OBUs get connected to internet via RSUs and HSs and use cellular radio networks such as GSM, Wi-Max and 4G when RSUs are absents.

Communication Architecture

Four categories are there in VANETs for communication types, which are closely related to components as described in Figure 2. This figure explains and illustrates the key functions of each communication type. The term "in-vehicle communication" refers to the in-vehicle domain, which is an increasingly vital component of research into VANETs. An in-vehicle communication system is essential for driver and public safety since it can monitor a vehicle's performance, notably driver fatigue and drowsiness. Drivers can exchange information and warnings via vehicle-to-vehicle (V2V) communication for providing support to the driver. Another category is

Figure 1. Different parts of VANET systems

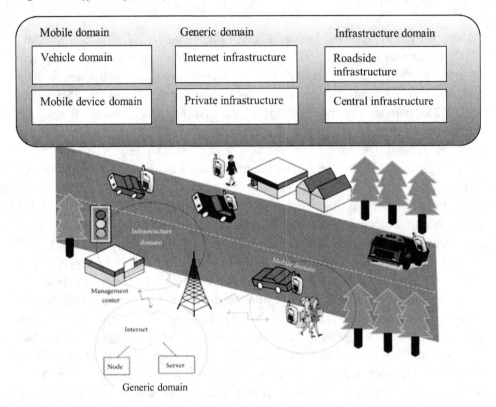

Generic domain

Vehicle-to-road infrastructure (V2I). In this type of connectivity, drivers can receive weather and traffic information in real-time, as well as environmental sensing and monitoring. In Vehicle-to-broadband cloud (V2B) communication, automobile interacts via a wireless broadband network for example 3G or 4G. The information of traffic and monitoring activity and infotainment is stored on the cloud. Due to this connectivity is imperative for providing assistance to driver and keeping track of vehicle.

Layered Architecture

The communication functions between nodes is divided into one of the seven logical levels as per the open systems interconnection (OSI) model. With this architecture, the session and presentation levels are dropped, and each layer can be further divided into sublayers. The design of VANETs can generally differ from region to region, which has an impact on the protocols and interfaces. A dedicated short-range communication (DSRC) was developed specifically for use in automobiles, along with a corresponding set of protocols and standards (Laberteaux & Hartenstein,2009). For DSRC communication, the US FCC has set aside 75 MHz of frequency between 5.850 GHz and 5.925 GHz. Some of the protocols are still being actively developed today, are intended to be used by the various layers.

Figure 2. Communication types in VANETs and their key functions

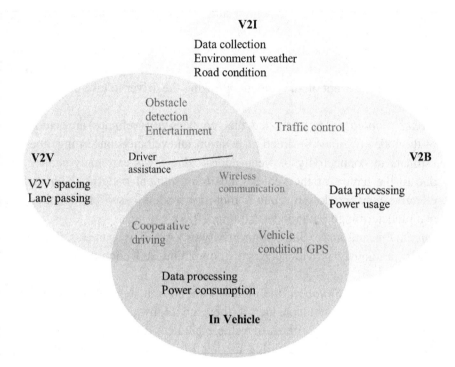

CHARACTERISTICS OF VANETS

Some distinguishing characteristics of VANET in the design of communication systems that plays an important role (Kaur et al.,2012, Chen,2015) are discussed here. These are:

- *Topology*: The topology of VANETs is dynamic and, is defined by the two primary parameters of speed and route selection, i.e., how quickly a vehicle changes its route. Assume two vehicles are moving at 90 km/h (25 m/s) apart and the link between them lasts 4 seconds (200 meters) if the transmission range is approximately 200 meters.

- *Localization Feature*: This feature transmits and sometimes determines a vehicle's geographical location, i.e., to and from one or more vehicles. Vehicle localization (estimation of location) is a global requirement. Vehicle usage is increasing in both developed and developing countries. The global vehicle fleet continues to grow on a daily and annual basis (Shinde et al.,2012, Almusaylim et al., 2020). Furthermore, the availability of position data (Davis, 2018) may enable applications such as driver safety, intrusion detection, inventory management, health monitoring, road traffic monitoring, and surveillance. Specific VANET applications, such as driver assistance systems, require a localization system with the highest accuracy, precision, and reliability. In these applications, the driver will receive information in advance in the event of an accident, allowing the driver to take precautionary measures.

- *Model of mobility:* The nodes and, their connectivity (vehicles) are determined by the node's (Vehicle's) direction, position, and velocity concerning time. The estimate of connectivity between nodes(vehicles) is extremely challenging due to the nature of the vehicle and its movement patterns. Nonetheless, effective network design requires mobility models based on vehicle speed and predetermined path models (Camp, 2015).

- *Infinite Power Source*: Since connectivity in VANET is seamless, the vehicle can continuously provide power to computing and communication nodes (vehicles).

- *Communication Situation*: the model of mobility is dependent on roadways patterns, they do not consider if they are highways, city or streets. So designing routing algorithms or vehicle prediction models need changes in roadways patterns. The mobility models for highways are simple and it is easier to perform prediction. On the other hand city mobility models are complex due to road structure. The obstacles like buildings, trees, signposts and other landmarks make predictions even more difficult.

- *Onboard Sensors for communication:* The onboard in any VANET are made up of several sensors. These sensors are capable in predicting a vehicle location and movement that provide effective routing and quality communication between vehicles.

- *Extremely Computational*: Active vehicles can communicate, sense, and compute in vehicular ad hoc networks.

- *Variable Network Size*: In vehicular ad hoc networks, the size of the network can change. It could be a country, a city, or a group of cities. Because of this, any protocol made for VANET is almost certainly ascendable.

- *Anonymous Addressee*: The most important thing for many VANET applications is to identify vehicles in a certain region, not the specific vehicle, which could help protect the privacy of nodes (vehicles).
- *Time-Dependent Data exchange*: sending packets of sensitive data that have time constraints is part of any VANET application. So it is not a good idea to compromise on performance.
- *Infrastructure Support*: A good infrastructure is backbone of a good Vehicle-to-vehicle communication and, so VANET is better than every other network.
- *Lots of energy and computers*: Because there is a lot of energy and computers, VANET has a lot of schemes that use a lot of resources, like the Elliptic Curve Digital Signature Algorithm (ECDSA) and Rivest–Shamir–Adleman (RSA) (Preet et al., 2017).
- *Better physical protection*: Because it's hard to break into VANET Nodes (vehicles), VANET services are safer than MANET services.
- *Network Partitions*: Because Vehicular ad hoc networks are always changing, there are often big gaps between vehicles in isolated clusters of nodes (sparsely occupied scenario).
- *Hard Delay Constraints*: VANET safety applications require on-time message delivery instead of hard data delay delivery in case of accidents.

ROUTING PROTOCOLS IN VANETS

Routing protocols evaluate each option before making a decision. the best for getting a point across. These paths are evaluated by the routing algorithms, which use them to select the best path leading to the destination, based on their connection, bandwidth, end-to-end delay, etc. For the sake of enhancing both road safety and passenger convenience, VANET routing involves sending data from the source vehicle to the destination vehicle over the network via a suitable path. To prepare for an impending thrust, the vehicles have access to data related to road safety (Da Cunha, 2014). High-dynamic topology and support for irregular connectivity are just two of the impressive features of this system. There are problems with inter-network communication in the initial VANET because of the lack of routing algorithms (Srivastava, 2020).

The best VANET routing protocol should include the following:

- Getting to Know Your Neighbors
- Their ability to transfer data
- Geographical information
- Predict where vehicles will be in the future

- Consider the uneven density of vehicles in the area

The existing VANETs are categorized into three distinct types of routing protocols, namely: broadcast, geocast, and unicast. The list of routing approaches and routing protocols is given in Table2.

- **Geocast/Broadcast**. The geocast/broadcast protocols are required in VANETs due to the requirement of disseminating messages to unidentified or unnamed destinations (Chen et al., 2011). Flooding is utilized for data transmission within a region or range. It is also feasible to send messages without being bombarded. Crash rates are reduced as parcel overhead is corrupted. Heading-based geocast for query distribution in VANET Inter-vehicle geocasting, distributed robust geocasting, and robust vehicular routing Timetable that is dynamic One of the sub routing protocols featured in Geo Cast routing protocols is the Geo Cast routing protocol. Daraghmi, 2013, reviewed the message broadcast protocols for VANETs, an interference aware routing scheme and spatially aware packet routing method. The interference aware routing equips the node with a multichannel radio interface which switches channels based on SIR evaluation and FROV. The FROV selects the retransmission. SADV finds the best path to forward the packet. This is possible by dividing the road into segments and choosing the farthest vehicle in the nonempty segment. Some other algorithms in this category are UMB, MHVB, MDDV, and V-Trade.
- **Multicast.** In driving situation such as crossroads, busy traffic, accidents, roadblocks and rough road conditions multicast is needed. In (Mchergui et al, 2020), the authors classify multicast protocols into two broad types. Topology-based approaches, such as GHM, ODMRP, and MAODV, produce group-based multicast trees and source-based multicast meshes using group addresses (which generates group-based multicast meshes). The location-based technique consists of RBM, IVG, LBM, SPBM and PBM. PBM uses the positions of all one-hop neighbors and all individual destinations respectively. The LBM utilizes a multicast area to keep destinations information for multicast packets. RMB and IVG main purpose is to handle safety warning messages by defining multicast protocols.

Table 2. Routing protocols applied in VANETs

Routing Method	Routing Protocols
Unicast	Ad hoc on-demand distance vector (AODV) General packet railways services (GPRS) Vehicle assisted data deliver (VADD) Dynamic source routing (DSR) A-START Position based multihop broadcast Improved greedy traffic aware routing Trajectory based data forwarding Connectivity aware routing Geographical source routing Opportunistic packet relaying in disconnected vehicular ad
Broadcast	Linkage protocol for highway automation (DOLPHIN) BROADcast COMMunication protocol Distributed vehicular broadcast (DV-CAST) Packet routing algorithm
Geocast	Distributed robust geocast protocol (DRG) Robust vehicular routing (ROVER)

- **Unicast.** Three types of unicast communication protocols are studied for VANETs:

  1. *Greedy:* A node chooses its farthest neighboring node and forwards packets to the destination node which is the same as improved greedy traffic-aware routing known as GyTAR (Chen et al., 2011).
  2. *Opportunistic:* Nodes wait for the opportunity to come, and then send the data to the destination on arrival similar to topology assist geo-opportunistic routing (TAGOR) (Daraghmi et al., 2013).
  3. *Trajectory-based:* Potential paths to the destination is identified by nodes. Further nodes determine the delivery of the data through nodes by applying trajectory-based data forwarding (Senouci et al., 2018).

ISSUES, CHALLENGES, AND FUTURE OF VANET

The challenges and issues that exist in the field of VANETs is discussed in this section. Also, some useful applications are described. Some issues in VANET are:

i.   Intermittent connectivity: Management of connections and control between cars and infrastructure is a significant challenge. Because of heavy vehicle movement or severe packet loss, intermittent connections in automotive networks are not considered good.

ii.   High mobility and location awareness: A high degree of mobility and location awareness will be required of the vehicles that participate in communication in future VANETs. Each vehicle in the network must know where the other vehicles are in order to respond to an emergency situation.

iii.   Management of heterogeneous smart cars: A large number of heterogeneous smart automobiles will be available in the future; the management of their irregular connections will be a challenge.

iv.   Security: Information about an individual's identity and whereabouts can never be completely protected. When cars talk to each other inside the infrastructure, they should be able to choose what information is exchanged and what information is kept private. Privacy can be guaranteed by analyzing sensitive data locally rather than sending it to the cloud.

v.   Network intelligence support: One of the future VANET's goals will be to support network intelligence. The edge cloud will collect and pre-process data from sensors on vehicles in future VANETs before it is shared with other parts of the network, such as regular cloud servers.

VANETs, in contrast to MANETs, have distinctive properties that necessitate the use of wireless communication technologies, different communication paradigms, and security and privacy measures (Dressler et al., 2011). Network connections, for instance, cannot be dependable over a long period. Researcher are looking for new ways to utilize existing infrastructure (e.g., Roadside units, cellular networks) for improving communication performance. Although specific VANETs challenges have been overcome, several significant research issues are still open (Khan et al., 2019). Some common technical and societal challenges faced by Vehicular ad-hoc networks are:

- Communication: When it comes to data exchange, there are three different methods such as Unicast, Multicast, and broadcast. All three are applied to implement the fully operational mode in VANET. The highly dynamic topologies, vehicle speed, bandwidth constraints, and homogeneous as well as heterogeneous network densities create challenges in VANET.

- Trustworthiness: VANETs depend on seamless connectivity, and an error-free environment to work efficiently in full operational mode. This is possible only when proper facilities like robust hardware and fault tolerant software are available. This is one of the key research areas in VANET.

- Privacy: data exchange between devices/vehicles is done anytime, anywhere and so different safety mechanisms are required. Some common mechanisms are authentication, digital signature, multifactor authentication, and cryptography.

- Cost allocation: These days, some automobiles are equipped with more expensive built-in VANET connection systems that not every vehicle owner can afford. A cost-effective VANET communication system design is therefore required to provide VANET services to every car user. Designing a VANET communication system with minimal resources could lower the cost of the vehicle.

- Quality of service: A important responsibility in the VANET is to provide Quality of Service (QoS) at a particular level. Due to the topology's great dynamicity, severe latency restrictions, unpredictable connectivity, etc., it is impossible to offer various users a higher level of service. As a result, it is one of the fascinating and challenging research fields in the creation of the VANET system (Shinde et al., 2021).

To solve the various issues in VANETs, some criteria are discussed below as a possibility in future VANETs.

- Low latency and real-time applications: Future VANETs must have low latency so that they can be used for real-time applications. Future VANETs should be able to support real-time applications like safety messages with very low latency.

- Great bandwidth: Shortly, there will be a lot of demand for high-quality video streaming and other entertainment and convenience apps. Also, traffic apps like 3D maps and navigation systems need to be updated automatically regularly.

- connectivity: In the future, connected cars will need to be able to talk to each other perfectly for VANETs to work. Fog devices must be always and very reliably connected to connected or self-driving vehicles. It must be able to keep communication systems from breaking down during transmission.

VANETS AND IOT APPLICATIONS

With the aid of IoT, the broadcasting protocol has been widely adopted in VANET for the distribution of data or messages for safety-based and non-safety-based applications to guarantee traffic efficiency (Mohamad et al.,2017). The idea of a smart city is crucial in today's technological era for enhancing the quality of services (QoS) provided to residents and, ultimately, their quality of life. Table 3 shows the

Table 3. Use of IoT in VANETs applications

SN	VANET Application	IoT application
1	Warning in case of collision	Smart cities
2	Assistance for Lanes	Environment monitoring
3	Accident detection and alert	Energy management
4	Traffic planning	Healthcare systems
5	Overtake warning	Building automation
6	Emergency warning	transportation
7	Point of interests allocation	Social network
8	Weather information	

connection or prerequisites between VANET and IoT. One may argue that the two are related.

Using a game-theoretical approach, IoT in VANET has also been used to make real-time decisions for IoT-based traffic light control. The authors proposed a way to control traffic flow or congestion at intersections (Bui et al., 2017, Avasthi et al., 2022). Their method can find priority vehicles that need help right away, like police cars, ambulances, and fire engines. This cuts down on how long these priority vehicles usually have to wait at intersections. Since cities are getting bigger every day, this kind of system is very important.

Safety application: The traditional objective of safety applications is to prevent accidents, which has resulted in the development of ad hoc networks for vehicles that communicate with other connected elements of the environment. These apps provide life-saving traffic assistance to drivers on the road. Driver assistance, alert information, and warning alert are its three subcategories.

Safety applications can be grouped as under:

- Real-time traffic: this data is collected at roadside units and shared with cars. This is an important part of fixing traffic jams and bottlenecks.
- Cooperative Message Transfer: Cars that are moving slowly or have stopped work together and send messages to other cars to avoid accidents by automating applications like an emergency.
- Post-crash notification: If a car is in an accident, it sends warnings to other cars so they don't get into the same kind of trouble.
- Road Hazard Control Change: The car in front of other cars can quickly tell them how the road is, what it is made of, or if there is a landslide.
- Traffic Watchfulness: The cameras set up at roadside units will help a lot to cut down on driving violations.

Non-safety applications: Non-safety uses are crucial for maintaining traffic efficiency and comfort while driving. Road users connected to VANET systems using IoT can benefit from weather information, location and current traffic movements on road networks, distance, Point of Interest (PoI) allocation, and social network (connected to the mobile network via smartphones) (Bauza et al.,2010). Commercial application is a popular category in non-safety applications. These involve value-added services (VASs) that emphasize improving passenger comfort (Eze et al., 2016). Commercial applications can be grouped as follows: -

- Remote Car: Through Personalization/Diagnostics, the download of customized vehicle settings or the upload of vehicle diagnostics from/to infrastructure is possible.
- Internet Access: Passengers can use the Roadside units (RSUs) of a VANET as routers to connect to the Internet.
- Downloading digital maps: Depending on the situation, a driver can download a local map for their own usage while visiting a new location.
- Real-time video relay: Vehicles can access and share or transfer multimedia files, including music, movies, news, e-lectures, e-books, etc.
- Value-added advertisements: the services given, such as shopping malls, gasoline pumps, gas stations, and highway restaurants, can advertise their services to all vehicles or passengers within range, even if the internet is unavailable.

Another type of non-safety application is a convenient application that provides convenience and comfort to the driver and manages traffic effectively making passenger's experience better. Convenience applications can be grouped under:

- Router Diversions: the driver can alter the route he or she intends to take in the event of traffic congestion.
- E-Toll: Electronic toll collection is an essential VANET application since it allows drivers to pay for tolls electronically at collection sites without having to stop.
- Availability of Parking: If a parking space becomes available, the vehicle will be alerted over the network when it is available.

TRUST MODEL AND ATTACKS IN VANETS

VANET's wireless ad hoc environment is shared and open, so some nodes may be hacked. A security solution must be able to identify potentially hazardous nodes and remove them from the network. Making the right security judgments and taking the appropriate security actions are considerably simpler when trust connections are transparent in real-time. It will be important to have a trust-based model that can manage nodes and evaluate their activities in a way that doesn't require a central authority. Based on the trust ratings they get, malicious nodes can be found, and the network can stop these nodes from taking part in any network communication. So, figuring out trust values is a very important part of making the network safer and more reliable. One entity's belief in another inside a network and certainty that the other would not act maliciously is defined as trust. An entity is any device that participates in communication. The first and second groups of the trust establishment paradigm are infrastructure-based trust and self-organized trust, respectively.

Attacks on VANETS Devices

Because of the enormous number of autonomous network members and the inclusion of the human component in VANETs, node misbehaviors in future vehicle networks are highly possible. Based on the layer used by the attacker, various attack types have been discovered and categorized (Balakrishnan et al., 2007). An attacker can disrupt a network at the physical and link layers by flooding the communication channel with unsolicited data. An attacker can also inject false messages and rebroadcast prior messages. On-Board Units (OBUs) can be tampered with or destroyed by some attackers (RSU). An attacker can insert fake routing messages or flood the system with routing information at the network layer. The common attacks in the network is described as follows:

1. False Information: The attacks are done by insiders who are smart and proactive. The attackers send incorrect information that can change how other drivers act. For example, an enemy could send incorrect information accident and a traffic jam on specific routes misguiding the traveler. This would send cars on different routes and free up a route for the enemy (Grover et al., 2013).
2. Using information from sensors to deceive: This is possible through a proactive and smart insider who uses this attack to change how the position, speed, and direction of other nodes are seen so that he won't be responsible for an accident.
3. Identifier Disclosure: An attacker is a passive and bad insider. It can keep track of a vehicle's path and use that information to find the vehicle.
4. Denial of service: A hacker may attempt to knock down the network by sending unnecessary messages over the channel. This kind of attack entails actions like channel blocking and delivering false messages.
5. Dropping Packets and Relaying Them: An attacker can drop packets that are meant to be sent. For example, an attacker could delete all alert messages that

were meant to warn vehicles that were getting close to the accident site. In the same way, an enemy can play back the packets after the event has happened to make it look like an accident.

6. Hidden vehicle: In this situation, intelligent vehicles aim to reduce wireless channel traffic. For instance, a car has alerted its neighbor's and is awaiting a response. When a vehicle receives a response, it recognizes that its neighbour is better equipped to convey the warning message to other nodes, therefore it stops sending the message to those nodes. This is due to the node's assumption that its neighbour will relay the message to other nodes. The system could be completely destroyed if this neighboring node is an enemy.

7. Attack of the Wormhole: A malicious node can record packets at one point in the network and tunnel them to another point using a shared private network with other malicious nodes. The attack will be worse if the malicious node just transmits control messages across the tunnel and no data packets (Grover et al., 2013).

8. Sybil Attack: The vehicle pretends to be more than one vehicle. These fake names also make it look like there are more cars on the road. The vehicle spoofs identities, the positions or details of the nodes on that network, this attack lets any type of attack happen.

Trust Model

Managing trust in vehicle ad hoc networks (VANETs) is one of the most difficult parts of setting up a safe VANET environment. Researchers have only made a few trust models to improve the reliability of information shared in vehicular networks. There are three main types of models shown: entity-oriented, data-oriented, and integrated trust models (Soleymani et al., 2015).

1. Entity-oriented trust model: The trust model of this kind focuses on how real cars are. Because vehicles move around a lot, trust models can't gather the important information about the sender and the vehicles around it that is needed to reach this goal. Now, the sociological trust model and the multidimensional trust management model are the most common entity-oriented models. The sociological trust model is based on a hypothetical principle of trust and confidence. It doesn't have a structure for finding different kinds of trust at the same time.

2. Data-oriented trust model: This model is different from the entity-oriented model that determines whether or not the participating nodes or cars can be trusted (Soleymani et al., 2015), the event-oriented model does not attempt to determine whether or not the participating nodes or vehicles can be trusted. The

objective in this trust model is to ensure that incoming messages are authentic. Various data-based trust models, including the RMCV intrusion trust model, the reputation-based trust model, the event-based reputation system, and the roadside-assisted data-centric trust establishment, have been proposed.

3.   Combine based trust: Uses a combination of individual node opinions and the opinions of other nodes to assess the authenticity of a message. In order for a message to be considered authentic, it must be accepted by the majority of cars. Beacon trust management approach [BTM] has been proposed (Raya et al., 2007) to address these issues.

MAINTAINING PRIVACY AND SECURITY OF VANETS

VANETs continue to have serious effects on privacy. While this is happening, the chance that your private information will be stolen and used against you grows as cars share thoughtful information about themselves and their neighbor's cars. Malicious or rogue vehicles are cars that do bad things like signal sniffing, pattern sniffing, changing packet information, throwing away packets, etc. So, several intrusion detection systems (IDS) approaches based on anomalies, signatures, watchdogs, cross-layers, and honeypots are shown. However, each has its limitations that can't stop invaders, adversaries, hostiles, and rogue entities from messing with normal network activities. The majority of researchers in the field have determined that security and privacy pose the greatest difficulty when transmitting data or messages over VANETs. Due to the online data integrity of the personal information act (Eze et al., 2016, Liang et al., 2015), kinematic data of VANET components cannot be exposed to a malicious server and user collusion (Talib, 2017).

   Recent research has addressed the issues of data ownership, massive data management, and legal culpability. Security architecture is essential for ensuring security and privacy to address the issue. Thus, the structure of a vehicular communication system should prioritize the provision of a communication scheme for safety-based applications, as it will generate a shared session key for secure network communication. Due to the poor security level of data clouds, roadside attackers may send bogus requests to cloud services for roads or parking, which causes confusion (He et al., 2014). Theoretical constraints and opportunities, the necessary IEEE standards, connectivity between vehicles and infrastructures, cross-layer design, mobility, validation, and cross-layer design are some other topics that have been briefly covered in VANETs study. Some security challenges are mentioned as follows:

- Validation: All of the messages that have been sent must be checked, starting with the first protest and moving on to the next. The central authority must make sure that every vehicle in the system is real.
- High Mobility: As the vehicle moves faster, its high mobility causes several problems, such as noise problems and the loss of handshakes. Because of this, the vehicles can't work together and communicate with each other safely.
- Area-Based Schemes: Beacon messages help us figure out where different vehicles are in a certain area. Sensors, GPS, and Laser, on the other hand, can be used to figure out exactly where the vehicles are.
- Real-Time System: Real-time systems can't be made in this area because people move around a lot. Because of this, it is hard to get alert messages to other devices in time before the deadline.

CONCLUSION

The goal of writing this chapter was to take a close look at the VANETs, IoT, and their applications. Future research in VANETs is now possible after comparison was drawn between the protocols currently in use and those that have recently been developed. Passenger safety, increased traffic efficiency, and entertainment are just a few of the many benefits that can be derived from using VANETs, or vehicle ad hoc networks. As technology advances and the number of smart vehicles rises, traditional VANETs are finding it increasingly difficult to deploy and manage due to a lack of flexibility, scalability, connectivity, and intelligence. To meet these demands, IoT technology is assisting VANETs in many ways in form of the Internet of Vehicles (IoV). In any case, VANETs of the next generation along with IoT will have unique requirements for autonomous vehicles with high mobility, low latency, real-time applications, and connectivity that conventional cloud computing may not be able to meet. The chapter provides an overview of VANET, the architecture of VANETS, Intelligent Transportation Systems, and the relevance of the Internet of Vehicles in implementing a better transport system for passenger safety and modern features.

REFERENCES

Ahmood, Z. (2020). *Connected Vehicles in the Internet of Things*. Springer Nature Switzerland AG. doi:10.1007/978-3-030-36167-9

Almusaylim, A. Z., & Jhanjhi, N. (2020). Comprehensive Review: Privacy Protection of User in Location-Aware Services of Mobile Cloud Computing. *Wireless Personal Communications*, *111*, 541–564. doi:10.100711277-019-06872-3

Avasthi, S., Chauhan, R., & Acharjya, D. P. (2022). Topic Modeling Techniques for Text Mining Over a Large-Scale Scientific and Biomedical Text Corpus. *International Journal of Ambient Computing and Intelligence*, *13*(1), 1–18. doi:10.4018/IJACI.293137

Avasthi, S., Sanwal, T., Sareen, P., & Tripathi, S. L. (2022). Augmenting Mental Healthcare with Artificial Intelligence, Machine Learning, and Challenges in Telemedicine. In Handbook of Research on Lifestyle Sustainability and Management Solutions Using AI, Big Data Analytics, and Visualization (pp. 75-90). IGI Global.

Balakrishnan, V., Varadharajan, V., Tupakula, U., & Lues, P. (2007). Team: Trust enhanced security architecture for mobile ad-hoc networks. *15th IEEE International Conference on Networks ICON 2007*, 182–187. 10.1109/ICON.2007.4444083

Baldessari, R., Bödekker, B., Deegener, M., Festag, A., Franz, W., Kellum, C. C., ... Zhang, W. (2007). *Car-2-car communication consortium-manifesto*. Academic Press.

Bauza, R., Gozalvez, J., & Sanchez-Soriano, J. (2010). Road traffic congestion detection through cooperative Vehicle-to-Vehicle communications. *Proceedings - Conference on Local Computer Networks*, 606–612. 10.1109/LCN.2010.5735780

Bui, K. H. N., Jung, J. E., & Camacho, D. (2017). Game theoretic approach on Real-time decision making for IoT-based traffic light control. *Concurrency and Computation*, *29*(11), e4077. doi:10.1002/cpe.4077

Camp, T., Boleng, J., & Davies, V. (n.d.). A survey of mobility models for ad hoc network research. *Wirel. Commun. Mob. Comput., 2*, 483–502. https://onlinelibrary.wiley.com/doi/10.1002/wcm.72

Chauhan, R., Avasthi, S., Alankar, B., & Kaur, H. (2021). Smart IoT Systems: Data Analytics, Secure Smart Home, and Challenges. In Transforming the Internet of Things for Next-Generation Smart Systems (pp. 100-119). IGI Global.

Chen, W. (2015). *A book on Vehicular Communications and Networks Architectures, Protocols, Operations and Deployment*. Elsevier.

Chen, W., Guha, R. K., Kwon, T. J., Lee, J., & Hsu, Y. Y. (2011). A survey and challenges in routing and data dissemination in vehicular ad hoc networks. *Wireless Communications and Mobile Computing*, *11*(7), 787–795. doi:10.1002/wcm.862

Da Cunha, F. D., Boukerche, A., Villas, L., Viana, A. C., & Loureiro, A. A. (2014). *Data communication in VANETs: a Survey, Challenges and Applications* (Research Report RR-8498). INRIA. https://hal.inria.fr/hal-00981126/ document

Daraghmi, Y. A., Yi, C. W., & Stojmenovic, I. (2013). Forwarding methods in data dissemination and routing protocols for vehicular ad hoc networks. *IEEE Network*, *27*(6), 74–79. doi:10.1109/MNET.2013.6678930

Davis, S. C., Diegel, S. W., & Boundy, R. G. (2018). *Transportation Energy Data Book* (36th ed.). Oak Ridge National Laboratory.

Dressler, F., Kargl, F., Ott, J., Tonguz, O. K., & Wischhof, L. (2011). Research challenges in intervehicular communication: Lessons of the 2010 Dagstuhl Seminar. *IEEE Communications Magazine*, *49*(5), 158–164. doi:10.1109/ MCOM.2011.5762813

Eze, Zhang, & Eze. (2016). *Advances in Vehicular Ad-Hoc Networks (VANETs): Challenges and Road-map for Future Development*. Academic Press.

Faezipour, M., Nourani, M., Saeed, A., & Addepalli, S. (2012). Progress and challenges in intelligent vehicle area networks. *Communications of the ACM*, *55*(2), 90–100. doi:10.1145/2076450.2076470

Grover, J., Gaur, M. S., & Laxmi, V. (2013). Trust establishment techniques in VANET. In *Wireless Networks and Security* (pp. 273–301). Springer. doi:10.1007/978-3-642-36169-2_8

Hartenstein, H., & Laberteaux, K. P. (2010). *VANET: Vehicular Applications and Inter-Networking Technologies*. Wiley Online Library. doi:10.1002/9780470740637

He, W., Yan, G., & Xu, L. D. (2014, May). Developing Vehicular Data Cloud Services in the IoT Environment. *IEEE Transactions on Industrial Informatics*, *10*(2).

Hossain, E., Chow, G., Leung, V. C. M., McLeod, R. D., Mišić, J., Wong, V. W. S., & Yang, O. (2010). Vehicular telematics over heterogeneous wireless networks: A survey. *Computer Communications*, *33*(7), 775–793. doi:10.1016/j.comcom.2009.12.010

Hussain, R., Rezaeifar, Z., Lee, Y. H., & Oh, H. (2015). Secure and privacy-aware traffic information as a service in VANET- based clouds. *Pervasive and Mobile Computing*, *24*, 194–209. doi:10.1016/j.pmcj.2015.07.007

Intelligent Transport Systems (ITS). (2012). *Framework for public mobile networks in cooperative its (c-its)s* (Tech. Rep.). European Telecommunications Standards Institute (ETSI).

Kaur, M., Kaur, S., & Singh, G. (2012). Vehicular ad hoc networks. *J. Glob. Res. Comput. Sci.*, *3*(3), 61–64. https://www.researchgate.net/publication/279973104_

Khan, U. A., & Lee, S. S. (2019). Multi-layer problems and solutions in VANETs: A review. *Electronics (Basel)*, *8*(2), 204. doi:10.3390/electronics8020204

Khelifi, A., Abu Talib, M., Nouichi, D., & Eltawil, M. S. (2019). Toward an Efficient Deployment of Open Source Software in the Internet of Vehicles Field. *Arabian Journal for Science and Engineering*, *44*(11), 8939–8961. doi:10.100713369-019-03870-2

Kumar, S., & Singh, J. (2020). Internet of Vehicles over VANETs: Smart and secure communication using IoT. *Scalable Computing: Practice and Experience*, *21*(3), 425–440. doi:10.12694cpe.v21i3.1741

Laberteaux, K., & Hartenstein, H. (Eds.). (2009). *VANET: Vehicular applications and inter-networking technologies*. John Wiley & Sons.

Lee, E.-K., Gerla, M., Pau, G., Lee, U., & Lim, J.-H. (2016, September). Internet of Vehicles: From intelligent grid to autonomous cars and vehicular fogs. *International Journal of Distributed Sensor Networks*, *12*(9). Advance online publication. doi:10.1177/1550147716665500

Liang, Li, Zhang, Wang, & Bie. (2015). Vehicular Ad Hoc Networks: Architectures, Research Issues, Methodologies, Challenges, and Trends. *International Journal of Distributed Sensor Networks, 5.*

Maier, M. W., Emery, D., & Hilliard, R. (2001). Software architecture: Introducing IEEE standard 1471. *Computer*, *34*(4), 107–109. doi:10.1109/2.917550

Maier, M. W., Emery, D., & Hilliard, R. (2002, August). 5.4. 3 ANSI/IEEE 1471 and Systems Engineering. In *INCOSE International Symposium* (Vol. 12, No. 1, pp. 798-805). 10.1002/j.2334-5837.2002.tb02541.x

Mchergui, A., Moulahi, T., Ben Othman, M. T., & Nasri, S. (2020). Enhancing VANETs broadcasting performance with mobility prediction for smart road. *Wireless Personal Communications*, *112*(3), 1629–1641. doi:10.100711277-020-07119-2

Mohamad, Abidin, Elias, & Zainol. (2017). *Optimizing Congestion Control for Non-Safety Messages in VANETs Using Taguchi Method*. Academic Press.

Patel, N. J., & Jhaveri, R. H. (2015). Trust based approaches for secure routing in VANET: A Survey. *Procedia Computer Science*, *45*, 592–601. doi:10.1016/j.procs.2015.03.112

Preet, E., Dhindsa, S. K., & Khanna, R. (2017). Encryption Based Authentication Schemes in Vehicular Ad Hoc Networks. *Conference: All India Seminar on Recent Innovations in Wireless Communication & Networking At: Baba Banda Singh Bahadur Engineering College Fateh garh Sahib*, 15–21.

Raya. (2007). *On data-centric trust establishment in ephemeral ad hoc networks* (Technical Report). LCA-REPORT-2007-003.

Schroth, C., Kosch, T., Strassberger, M., & Bechler, M. (2012). *Automotive internetworking*. John Wiley & Sons.

Senouci, O., Aliouat, Z., & Harous, S. (2018). A review of routing protocols in internet of vehicles and their challenges. *Sensor Review*.

Shinde, S. S., & Patil, S. P. (2010, July–December). Various Issues in Vehicular Ad Hoc Networks: A Survey. *Int. J. Comput. Sci. Commun.*, *1*(2), 399–403.

Shinde, S. S., Yadahalli, R. M., & Shabadkar, R. (2021). Cloud and IoT-Based Vehicular Ad Hoc Networks (VANET). *Cloud and IoT-Based Vehicular Ad Hoc Networks*, 67-82.

Shinde, S. S., Yadahalli, R. M., & Tamboli, A. S. (2012, April–June). Vehicular Ad Hoc Network Localization Techniques: A Review. *Int. J. Electron. Commun. Technol.*, *3*(2), 82–86.

Soleymani, A., Abdullah, A. H., Hassan, W. H., Anisi, M. H., Goudarzi, S., Rezazadeh Baee, M. A., & Mandala, S. (2015). Trust Management in Vehicular Ad-hoc Network. *EURASIP Journal on Wireless Communications and Networking*, *2015*(1), 146. doi:10.118613638-015-0353-y

Srivastava, A., Prakash, A., & Tripathi, R. (2020). Location based routing protocols in VANET: Issues and existing solutions. *Vehicular Communications*, *23*, 100231. doi:10.1016/j.vehcom.2020.100231

Talib, Hussin, & Hassan. (2017). Converging VANET with Vehicular Cloud Networks to reduce the Traffic Congestions: A review. *International Journal of Applied Engineering Research*, *12*(21), 10646-10654.

Tanuja, K., Sushma, T.M., Bharati, M., & Arun, K.H. (2015). A Survey on VANET Technologies. *Int. J. Comput. Appl.*, *121*(18), 1–9.

Zeadally, S., Chen, Y.-S., Irwin, A., & Hassan, A. (2012). Vehicular ad hoc networks (VANETS): Status, results, and challenges. *Telecommun. Syst.*, *50*, 217–241. https://link.springer.com/10.1007/s11235-010-9400-5

Chapter 2
Industrial Automation and Its Impact on Manufacturing Industries

Subha Karumban
Alagappa University, India

Shouvik Sanyal
Dhofar University, Oman

Madan Mohan Laddunuri
Malla Reddy University, India

Vijayan Dhanasingh Sivalinga
 https://orcid.org/0000-0003-0128-5397
Arupadai Veedu Institute of Technology, India

Vidhya Shanmugam
REVA University, India

Vijay Bose
Vaagdevi College of Engineering, India

Mahesh B. N.
Reva University, Bengaluru, India

Ramakrishna Narasimhaiah
School of Humanities and Social Sciences, Jain Deemed to be University, Bengaluru, India

Dhanabalan Thangam
 https://orcid.org/0000-0003-1253-3587
Acharya Institute of Graduate Studies, Bengaluru, India

Satheesh Pandian Murugan
Arumugam Pillai Seethai Ammal College, India

ABSTRACT

Industries in recent days has been facing several issues including a dearth of labor, moribund joblessness rates, and towering labor turnover. This is projected to increase further in the upcoming days with the rise of the aged population. All these challenges can be addressed by the technology called industry automation. But there is an acuity that industrial automation will lead to job losses. However, in ingenious automation technology, lots of positive outputs can be achieved such as

DOI: 10.4018/978-1-6684-4991-2.ch002

higher productivity and enhancement in yield. The role of automation in health and safety is awesome, but the brunt of this technology on jobs and efficiency worldwide has not been studied fully. In addition, automation technology safeguards the workers from highly dangerous work zones such as mines, space research, and underwater research. Thus, industrial automation is ready to serve humankind and business in various ways. This has been explained in this chapter.

INTRODUCTION

The ultimate objective of any business is to maximize the profit by minimizing the operating cost and every organization strives for the same; otherwise, it would go out of the business. On the other side manufacturing industries are facing lots of challenges right from the shortage of workforce, high labor turnover, mishaps, and injuries. A yearly industry research statement reveals that the manufacturing industries have to face two most important challenges, one is fulfilling customers' willingness and the other is answering the concern over the stern shortage of the required skills. According to the report it is also expected that around 3.5 million industrialized jobs will require to be packed and the deficiency would be almost 2 million by 2025 (Scott Technology Ltd, 2019). In the meantime, the haste of advancements in technology and the availability of more alternatives have stimulated consumers' expectations to rise at a brisk rate. Moreover, customers in each business wish to obtain good quality products, quicker delivery, better customer service and all these should be available while paying comparatively less for these services. Countries like North America, South Korea, Europe, and Japan and industries like meat and food processing and fruit growing are gotten pretentious more than other sectors due to the shortage of labor. As a consequence, automation technology is started to use progressively more in the production and manufacturing process. Current chapter sum-up the outcomes of the latest investigations in a comprehensive manner, and provides some real-world paradigms, relevance, and impact of industrial automation. Further, the chapter has designed to inform the challenges faced by the manufacturing industries due to labor shortage, the advantages of industrial automation, and the initiatives required to take to implement the automation technology to transform the manufacturing sector into a successful one. The other parts of this chapter explain the present labor shortage situation worldwide and estimation of the aging populations, automation and its brunt on jobs, efficiency, and profitability, succumb and excellence, health and safety of the workforce, and the initiatives required for promoting automation across industries (Nigel wright, 2021).

LABOUR SHORTAGE, HIGH STAFF
TURNOVER AND POPULATION AGING

Rising population and demand for goods and services kicking the manufacturing industries to increase the production capacity to meet material demand. Meantime, manufacturing industries are facing the issue of labor shortage, and with the available labor force, they cannot cope with required consumption. Several regions and manufacturing industries in the world are pretentiously worsening than others. From 2018 onwards, the manufacturing industries are facing the labor shortage issue worldwide. It is due to a lack of qualified workers with appropriate technical skills, rising retirement and aging population, mounting difficulties in the international supply chain, and academe. As a result, the labor shortage in the worldwide manufacturing sector worsening further and it may go up to 80 lakh by 2020, resulting in 607 billion USD worth of income less likely to happen. This could worsen further in the countries which are facing the issue of labor shortages already. In the case of Hon Kong, the labor shortage would rise to 80 percent in the manufacturing sector within the next 20 years (Terry Brown, 2021).

As far as Germany is concerned, about 1.5 million workforces are required to the manufacturing sector, but there is no workforce availability, and it paves way for automation of industries. Japan has witnessed around a 15percent labor force decline in its overall availability of the workforce in the manufacturing sector since 1970 and is expected to 5 percent revenue loss by 2030 because of its labor shortage in the manufacturing industries, and it would equate to 10 billion USD (Autor, 2015). According to Manufacturing Institute and Deloitte's report, there will be a demand for 5 million workforces to fulfill the vaccines in the USA's manufacturing sector, however, 2.5 million workforces only can attain and the rest of the vacancies cannot be fulfilled by the reason of lack of skillful or competent workforce. 2.5 million Labor shortage is a considerable number because the USA contributes around 18 percent in the worldwide manufacturing production and it also contributes 12 percent approximately towards the national output. On average around 21,500 jobs were created every month in the USA manufacturing sector in 2018. whereas the number has come down to 5750 per month in 2019. Further, the manufacturers have stopped recruitment due to the non-availability of the young workforce after the economic recession; as a result, the USA manufacturing sector is facing high baby boomer retirements. Economic experts have foreseen a declining tendency in the global production and manufacturing industry due to a shortage of labor, and it will prolong at least for the first six months in 2020. As a result, the GDPs are being affected by the shortage of labor in the manufacturing industries worldwide (Nolan Schroeder, 2021). The blend of moribund joblessness rates in a moribund job market leads to severe problems and challenges. Because, when there is an inadequate workforce,

Figure 1. Number of aged population by various country groups from 1980 to 2050.
Source: United Nations, world Aging Population 2017 highlights

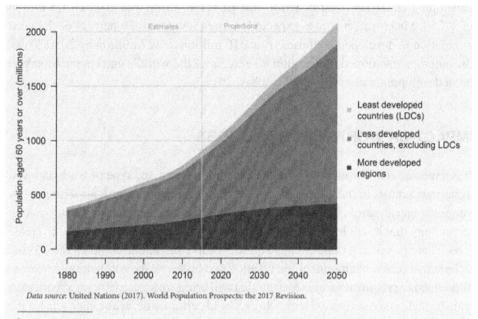

Data source: United Nations (2017). World Population Prospects: the 2017 Revision.

[1] Following common practice, the "developed regions" include Europe and Northern America plus Australia, New Zealand and Japan, while the "developing regions" include all other parts of the world. The use of these terms in the present report does not imply any judgement as to the current developmental stage of a particular country or region.

the demand for workers may augment in some industries, and it may lead to labor turnover in some other industries, and it may be the custom in the industries. The labor turnover has augmented from 1.5 percent to 2.5 percent per week in the case of the USA's meat and poultry processing industry and it is happening never before in these sectors. If these industries are trying to adjust this turnover by the labor replacement, they have to incur 4.2 percent additional cost in the overall processing cost, resultant it may lead to the profit loss worth of hundred million. This labor shortage and turnover will be increased further in manufacturing industries in the upcoming days owing to the aging population (DESA, 20117). Because the global aged population of more than 6-0 years was 960 million in the year 2017, and it was 380 million in 1980 and is also expected to increase double as 2.5 billion in 2050. It can be understood easily by looking into figure1.

It is also important to note that the growth rate of the aged population is more in developing countries than in developed countries. As a result, the developing countries are being the domicile to the world's large, aged population. The developing countries were domicile for 55 percent of the world's aged population who crossed

60 plus in the year 1980. Further, the world's two-thirds of aged populations were lived in developing countries as per 2017 statistics. The aged population in developing countries is estimated to rise from 650 million in 2017 to 1.5 billion in 20250. Meantime, it is also expected to increase around 36 percent of the aged population in developed countries from 310 million to 427 million by 2050. Thus, the outcrops mentioned that around 80 percent of the world's aged population will be in developing nations by 2050 (DESA, 20117).

IMPACT ON WORKFORCE AND JOBS

It is expected that automation would change the pattern and type of work and jobs in various sectors in the upcoming days, and this has been experienced already by various countries across the world. However, there are statements about the automation technology that it will lead to massive job loss, though it has the potential to create jobs. There is research done already regarding this confrontation and all those bring to light that different amount of job creation and losses by the automation, however, situations are reliant on various reasons that will bring constructive or unconstructive transformations. Automation technology would certainly be an alternate one for the manual workforce since it is ready anticipated to do all the work. Various experts have exaggerated the scope of automation and its substitution for manual labor, but, they have forgotten to mention the various complementarities existing in between automation technology and manual labor that enhance productivity, ensure safety in the hazardous workplace, generate income, and enlarge labor requirement (McKinsey Global Institute, 2021). In supporting this statement, (PWC, 2021) also mentioned that the so-called automation technologies will generate various patterns and types of jobs. Of which some may have a direct relationship with automation technologies, however, most of the industries prefer this automation technology as it helps to increase the output, generate incomes, and promote the wealth of the organization. While these additional incomes are invested in new avenues, new jobs would be created, and it may require additional labor for the same. The discussion is here not about the constructive or destructive aspects of the automation technology, but about the extent to which it has made a brunt on the works. However, there is no clarity in that, and it has been cleared by the various studies they have presented in Figure 2.

The discussion here is not whether automation has a constructive or unconstructive brunt on jobs; but about the extent to which automation impacts the jobs. This lack of lucidity is confused different stakeholders, and the same has cleared by (Nigel wright, 2021) by analyzing various reports published by various agencies between 2012-13 and 2017- 18, and small uniformity or conformity lives on a total number of jobs will be gained or lost through automation, globally, in the upcoming years. As

Figure 2. Prediction of Jobs Gained and Loss by the Automation Source: /itchronicles. com

Predicted Jobs Automation Will Create and Destroy

When	Where	Jobs Destroyed	Jobs Created	Predictor
2016	worldwide		900,000 to 1,500,000	Metra Martech
2018	US jobs	13,852,530*	3,078,340*	Forrester
2020	worldwide		1,000,000-2,000,000	Metra Martech
2020	worldwide	1,800,000	2,300,000	Gartner
2020	sampling of 15 countries	7,100,000	2,000,000	World Economic Forum (WEF)
2021	worldwide		1,900,000-3,500,000	The International Federation of Robotics
2021	US jobs	9,108,900*		Forrester
2022	worldwide	1,000,000,000		Thomas Frey
2025	US jobs	24,186,240*	13,604,760*	Forrester
2025	US jobs	3,400,000		ScienceAlert
2027	US jobs	24,700,000	14,900,000	Forrester
2030	worldwide	2,000,000,000		Thomas Frey
2030	worldwide	400,000,000-800,000,000	555,000,000-890,000,000	McKinsey
2030	US jobs	58,164,320*		PWC
2035	US jobs	80,000,000		Bank of England
2035	UK jobs	15,000,000		Bank of England
No Date	US jobs	13,594,320*		OECD
No Date	UK jobs	13,700,000		IPPR

mentioned in the table, the forecasting results of various studies are closer to 2020-21, a decade ahead the figures have become unintelligibly larger. While seeing the figure of 1.8 million, it looks like a small portion of the jobs will go, however after two years, and a decade after the total number has increased more than 0.5 million, and it is very hard to digest. Winick (2021) Noted that these figures are calculated by various international experts in the field of technology automation and economics as well, and they have concluded that it is very difficult to predict the total number of jobs that will be lost instead of industry automation. To support this result, (Frey and Osborne, 2013) mentioned in their report that by 2030 around 2 billion works would be automated, McKinsey Global Institute (2021) report revealed that around 400 to 800 million works would be automated during the same period, however, Graetz (2021) conducted a study in various countries during 1996 to 2012 and cited in his report that job loss due to the application of industrial robotics is not authentic information, instead, it would create millions of job. Correspondingly, several reports (Gartner, 2021; IFR, 2021; McKinsey Global Institute, 2021) forecasted the job opportunities generated by automation technology globally. It is expected around 2 million to 890 million works would be automated in developed countries such as the USA, Europe, Japan, and Korea. To support this statement PWC report (2021) revealed that around 55,000 robots have been employed in industrial operations from 2010 to 2016. Further, 262,600 works were generated in the sector during the mentioned timeline. While the German manufacturing sector employed 300 industry robots to 10,000 employees, however, it also has generated around 72,317 job opportunities from 2010 to 2016 (Graetz, 2021).

INDUSTRIAL AUTOMATION ON THE PRODUCTIVITY AND PROFITABILITY

The growing importance of industrial automation in the manufacturing industry employs robotic labor and thereby it is replacing the roles played the human labor. Its result this shift leads to drastically augmented ineffectiveness of manufacturing industries. Since the Machines never take a rest like human labor, the work is happening more rapidly. Thereby automation enables the manufacturing industries to maximize their output with greater speed, without an additional workforce. Thereby industries can realize faster revolving times, and can also reduce the wait times in between projects. Thus, automation is very much increasing the competency of the industries to manufacture more products at quicker rates. Though industrial automation helps to maximize the output, it is requiring huge investment during the initial time. However, it will help the organizations to cut down the costs over a longer period of time. Industrial automation helps to get better productivity and profit, further automation technology performs the jobs more accurately, efficiently and maintains the quality of outputs constantly than manual laborers. Centre for Economics and Business Research (2021) conducted a study to assess the impact of industrial automation on industrial productivity and profitability in 17 countries during 1993 and 2007. The results of the study revealed that automation technologies increased the GDP of the studied countries up to 0.37 percent and raise the productivity of the industries by around 0.37 correspondingly. These statistics reflect 12 percent growth in the total GDP and 18 percent growth in the labor productivity of the 17 countries during the mentioned period. Automation and its economic impacts on 27 OECD countries have been studied by (Berg, Buffie, & Zanna, 2021) from 1993 to 2015, and the results of the study reveal that there is a constructive relationship between industrial automation and productivity. Increasing one unit of automation lead to 0.04 percent productivity of the labor. Moreover, industry automation also leads to economic progress resultant adding one percent of automation induces 0.03 percent of GDP growth. McKinsey Global Institute (2021) predicted the growth possibilities of 0.8 to 1.5percent ensured by the industry automation annually. Kowitt (2021) mentioned that a small change in the amount of industrial automation will lead to a massive rise in the output, moreover, industry automation and humans are interrelated with each other.

YIELD, SPEED AND QUALITY

Improvement in operational speed is one more advantage of industrial automation. McKinsey Global Institute (2021) also explained that the speed of production is an advantage of automation and its benefits would be varying from different applications. A hydroponic-based Bowery farm from New Jersey mentioned that the productivity of this type of automated farm is more than regular farms (Greenleaf Enterprise, 2021); since it is indoor-based as well as automated. Moreover, this type of automated farmlands is controlled by sensor-enabled technology, and thereby the hotness and wetness are maintained properly, thus it induces a high growth rate, higher crop succession, and yield per cultivation (Greenleaf Enterprise, 2021). This type of farming helps to save water, control insects, avoid pesticides and fertilizers; it also helps to curtail wastage of the water and thus it improves crop aroma (Levert & Héry 2019). Uguina and Ruiz (2019) found that automation in the pork boning industry reduces wastages and increases the volume of primal cuts. With the help of automation, these entities save the yields roughly 0.41 percent carcass, for three types of animals, and it was equal to AU$3.065 of profits regularly. Apart from the profit increment, the wasted meat can be converted to consumption. If the automated pork boning unit was utilized in the Australian pork processing industry, it could find an extra yield of 4,510 tonnes of porks on the feast slab in the year 2017. As industrial automation technology supports the manufacturing processes with the programmed machines by computer algorithms, possibilities for human error in the production process can be avoided. In addition to this, there are technologies in recent times with the novel algorithms such as machine learning allowing the machines to learn the things necessary for the smooth running of the industry. Machines themselves rectify the errors and make suitable modifications to avert possible errors in the near future. Even the recent time machines have the capacity to make decisions in difficult times, by recognizing existing defects, thereby it ensures the quality of the production process, and also ensuring prognostic and adaptive protection procedures. Thus industrial automation along with machine learning allows industries to get better production processes through more accurate actions with no human intervention.

WORKERS WELLNESS AND PROTECTION

The role played by automation in safeguarding the health and safety of the employees is excellent. Murashov, Hearl, and Howard (2015) revealed in their study that industrial automation is an important one for increasing the workload on one side; on the other side, it lessens the physical efforts of the labor and repetitive tasks. As per the report, the increased workload be the cause for an augment in job-related occurrences, this hypothetical information is not suited to the current statistics. Apart from the statistics, industry automation technologies are playing a major role in the most dangerous work zones and be an alternative to the places where repetitive tasks need to be done. Manual welding is having more chances for the accidents in the workplace such as electric shocks, smoke, eye problems, wounding to the body, cuts, and so on. But these kinds of problems can be avoided with the help of automatic welding technology (ISO, 2011). Automation technology is taking care of the employees by removing them from a hazardous workplace, thus it ensures the well-being of the workers in the workplace. According to the USA's Department of Labor's Occupational Safety and Health Administration, on average in a year the USA's industries are spending around 175 USDs due to the wounding and accidents caused by the improper working environment. Because of this reason, more manufacturing industries started to implement automation technologies in their working premises to reduce accidents, ensure workers' safety; thereby 25 to 50 percent of the costs can be saved by automated technologies (Bessen, 2021).

TRANSITION TOWARDS PRODUCTION AUTOMATION

Even though industrial automation comes up with lots of benefits, it is not free from the threat that loss of jobs. Of course, the industrial automation processes make obsolete some types of jobs, it is not eliminating jobs, and instead, industry automation is generating a new set of jobs. As a result, there will be a need for manpower where the machines cannot perform some of the things such as instinct, inventiveness, and care. In many cases tasks are automated and are cyclical, spraining, and time intensive. In such cases allocating this type of task to machines let the manufacturers generate more employment roles for the association, communication, and decision making. The finest way to evade job cutting in manufacturing industries is to concentrate more on providing training on the latest updates. Thus, industrial automation will lead to developing a well-organized manufacturing process and help to achieve augmented profits. Hence workers need not worry about losing jobs or shifting to other industries, they just need to be ready to train themselves to perform machine collaboration and decision-making related roles in the industry (Leslie, 2020). As industries are adopting the automation process gradually, they may regularly retrain workers to engage with these new positions. Furthermore, each industry will take individual assessments about which type of processes will be helpful, most competent, and gainful to automation, and accordingly the choices will be given to the workers. It is understood from the depicted information that, there is harmony with various researchers and industry stakeholders (Autor, 2015; Massimiliano & Presidente, 2021; PWC, 2021; McKinsey Global Institute, 2021) they have ensured the industrial transformation from traditional manufacturing to automated modern manufacturing setup, for the same there should be a harmonized efforts among governments, policymakers, industry experts, economists, as well as companies. As per cited experts' opinion, Governments of every country should work closely with the industrial sector of the country to discover what kind of jobs need to be developed or redesign by the industrial automation technology for the same a well- designed curriculum should be framed and implemented in the educational institutions and the suitable training programs should be given to the workers to update themselves as per the requirement of the modern technology or jobs (Lee & Anderson, 2021). Governments are supposed to think about incentivizing enduring education through financial support for providing skill development training programs. Governments should also think about the policy implementation to promote early adoption of industrial automation and increase investment in industrial automation technologies and ICT-enabled industrial infrastructure desirable to maintain successful industry automation. Industries should appraise their organizational actions to evaluate and extract possible value from industrial automation with a strategic plan that contains both investment and capital. Industries should invest more in the training programs to augment their reserve of elusive assets which will help to attain continuous growth. The academic sector should try to modify the curricula and stress more on science, technology, engineering, and mathematics (STEM) to improve the employability skills but also endorse the individual skills that automation technologies will not reinstate various qualities creativeness, understanding, critical thinking, etc (Craig & Ivanov, 2020).

INDUSTRY AUTOMATION IN THE DEVELOPED ECONOMIES

PWC study revealed that around 30 percent of jobs will be influenced by industrial automation by the 2030s in the United Kingdom (UK) and its impact will be reflected in various industries and jobs. Manufacturing industries, transportation sector, and logistics and supply chain management industries are having the possibility to automate their operations by 2030 with 45 percent, 52 percent, and 39 percent respectively. However, this industrial automation will not replace human labor with robotics, instead, it will go with the concept of cobots. It means industries will be having collaboration with robots thus changing the landscape of the manufacturing industries (Guilherme Luz, Rossini, Costa, Staudacher & Sawhney, 2021). By promoting organizational competency without affecting employment or the bottom line of the organization, industrial automation can really increase the output of the organization. Thus, on average UK industry at present incorporates 75 robotic units per 10,000 workers with cobots concept and transforms the functioning of manufacturing industries. However, the average use of robots in the manufacturing industry has increased from 75 units to 113 per 10,000 workers and becoming a new global record. Western European countries employ 225 units, and Nordic European countries incorporate 200 units, in their production process. As far as North America is concerned it has been installing 155 units, Southeast Asia using 125 units. Figure 3 explains the world's top ten countries using automation technology in their production process (IFR Press Room, 2021).

According to the IFR International Federation of Robotics report 2019, Singapore is the topmost country in using robotics in the manufacturing process and it is using 918 robotics per 10,000 workers. The electrical and electronics industry is the primary customer of using industrial robotics in Singapore with shares of 75 percent of the total operations. Followed by Singapore, South Korea is in the second position by utilizing 868 robotics per 10,000 workers. As Korea is good at manufacturing LCD and memory chips, more robotics are used here, and it is also a most important producer of automobiles and batteries for electric cars, majority of the robotics used here too. Japan is in third place by using 364 units, Germany in fourth place by installing 346 units. Japan is the largest manufacturer of robotics in

Figure 3. The world´s top 10 Most Automated Countries. Source: IFR Press Room

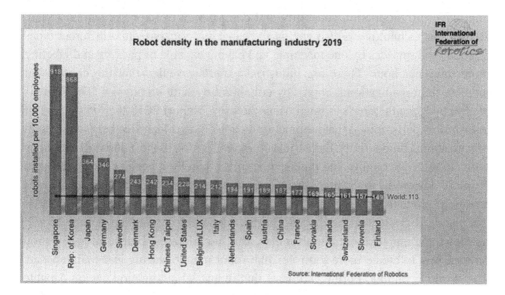

the world, where 45 percent of the worldwide robot production is made in Nippon (foundry4.com, 2019). The electronics industry utilizes 35 percent, 32 percent of the robotics have been used by the automotive industry, and 13 percent utilized by the metal manufacturing industry. In Germany robotics concentrated more in the automobile industry and it is the highest user of robotics in the world. This sector also ensures employment opportunities to 720,000 workers in 2010 and this figure has increased to 850,000 in 2019. Sweden is in the fifth position by adopting 274 robotics in the business operations with a split of 35 percent of robotics used by the metal industry and the remaining 35 percent used by the automobile industry. The utilization of robotics in the United States of America (USA) augmented to 228 units. The utilization of robots in Chinese industries increasing animatedly, as a result, China is the 15th largest country in the world using robotics in the manufacturing industries (Maximiliano & Asha, 2019).

BENEFITS OF INVESTING IN industry AUTOMATION

Incorporating automation technologies in the manufacturing industries helps to increase the organizational output by 24x7 in a year with no stoppage or command. Moreover, robotics will execute repetitive tasks without errors; handle the dangerous work zones carefully with no accidents and collapse. Thereby automation technologies help to ensure the safety and health of the employees in the workplace. On the other side, automation technologies also help to reduce the cost of production by improving the production capability and competency of the organization (Niall, 2020). This technology is also having the capacity to manufacture different types of goods and services by one machine, and having capable of packing and labeling more units per hour. There are industrial experts revealed that the cobots will increase the organizational output by collaborating with employees. The same is proved by Manufacturer's Annual Manufacturing Report 2018 that 95 percent of industrial experts opined that automation-enabled smart factories help to augment organizational productivity. Industrial automation performs the industrial operations with cutting edge quality and precise measures, thereby it enables the business to be more advanced, innovative, and competent, and thus support customers with advanced products and services (Megan Ray, 2020). However, in the majority of the places, it could see the discussion that industrial automation definitely will cut down the employment opportunities. But in real cases, Industrial automation paves a way to create more job roles like robot manufacturing and maintenance. The World Economic Forum estimates that industrial automation having capable of creating a job for 133 million people worldwide by 2030. Hence instead of fear of losing employment opportunities to industrial automation, it is better to invest more in skill development and retraining. Thereby both employees and automation technologies collaborate in the industrial automation process using cobots. Moreover, the cobots don't need a wider knowledge and skills on various aspects such as programming; however automation-related training tasks are time-consuming. With the help of retraining and skill development programs, industries can realize a lot of benefits along with automation technologies. It's also imperative to understand how automation can brunt the overall economy. For the same, the PwC has revealed that industrial automation has the potential to boost the worldwide GDP from investing in artificial intelligence and it could be, around, £12.5 trillion by 2030 (Gustavo Sepulveda, 2020).

IMPLICATIONS

Due to the industry 4.0 evolutions, there is a debate among various stakeholders regarding position jobs and employment in the years to come. However, all the countries in the world turned towards industrial automation due to various reasons such as shortage of labor, aged population, higher staff turnover, higher wage, and so on. Since automation has lots of benefits such as increasing productivity, producing consistent quality products, ensuring the health and safety of the workers, and so on. Much research works also indicate that industrial automation is a supportive technology that increases the productivity and output, rather than alternate for, manual workforce and thus improves the quality of the work and output fulfilling the tasks with perfection. Though, automation technology causes unconstructive socio-economic brunt in the short term, discontinuing or avoiding automation will not be a feasible solution. The competitiveness of any business or industry can be strengthening by increasing the usage of automation technology. There is distress about the reduction in middle- level jobs and income and rising wage disparity is justifiable, but all these issues are not raised by automation. However, industry automation stimulates the demand for high-skilled and income workers, its shock on low- skilled and income job openings will be clear slightly. Industry automation will redesign the way of work in the years to come, with massive possibilities for developments in output, augmented nationwide competency and the enhanced quality outputs, and higher compensation, with a sophisticated work environment. Hence, Governments and industrial sectors should work together to make a conducive atmosphere that will allow the workers, industries, and countries to obtain the benefits of this technology. Further, to support the research and development in the industrial automation and robotics sector there should be reasonable financial support ensured by the government. Also, there should a training program for the employees to improve their technical know-how in accordance with the technology up-gradation. Hence, there should be a group effort between the industrial sector and governments to make this industrial transformation safe, smooth, and useful to various stakeholders.

CONCLUSION

Technology is advancing at a rapid rate, causing drastic changes across most industries. Manufacturing is no different, with automation taking over the roles that humans once had, making certain jobs obsolete. The best thing that manufacturers can do to stay relevant in the modern age is to embrace the change. Invest in automation and new technology that will ultimately lead to better and smarter manufacturing processes. The more manufacturers focus on the future, the more concerned they'll be with retraining employees to prepare them for new types of roles in the modern age. Human labor is not becoming obsolete in the technology age. In fact, it's even more important than ever. Smart manufacturers are investing in automation for mundane tasks and more complex roles for their employees to stay relevant in the modern age.

REFERENCES

Autor, D. (2015). Why Are There Still So Many Jobs? The History and Future of Workplace Automation. *The Journal of Economic Perspectives*, 29(3), 3–30. doi:10.1257/jep.29.3.3

Berg, A., Buffie, E., & Zanna, L. (2018). *Should We Fear the Robot Revolution? (The Correct Answer is yes)*. https://www.imf.org/en/Publications/WP/Issues/2018/05/21/Should-We-Fear-the-Robot-Revolution-The-Correct-Answer-isYes-44923

Bessen, J. (2016). Computer don't kill jobs but do increase inequality. *Harvard Business Review*. https://hbr.org/2016/03/computers-dont-kill-jobs-but-do-increase-inequality

Bloombery. (n.d.). *This High-Tech Farmer Grows Kale in a Factory*. https://www.youtube.com/watch?v=AGcYApKfHuY

Brown, T. (2021). *The Impact of Automation on Employment*. https://itchronicles.com/automation/the-impact-of- automation-on-employment/

Calì, M., & Presidente, G. (2021). *Automation and Manufacturing Performance in a Developing Country*. Academic Press.

Centre for Economics and Business Research. (2017). *The impact of automation – a report for Redwood*. https://cebr.com/wp/wp-content/uploads/2017/03/Impact_of_automation_report_23_01_2017_FINAL.pdf.

foundry4.com. (2019). https://foundry4.com/5-countries-with-the-most-robots-in-manufacturing

Frey, C. B., & Osborne, M. A. (2013). The Future of Employment: How Susceptible is Jobs to Computerisation? University of Oxford.

Gartner. (n.d.). *Gartner Says By 2020, Artificial Intelligence Will Create More Jobs Than It Eliminates*. https://www.gartner.com/newsroom/id/3837763

Graetz, G., & Michaels, G. (n.d.). *Robots at Work*. CEP Discussion Paper No 1335, Centre for Economic Performance, the London School of Economics and Political Science. cep.lse.ac.uk. https://cep.lse.ac.uk/pubs/download/dp1335.pdf

Graetz, G., & Michaels, G. (n.d.). *CEP Discussion Paper No 1335 - Robots at Work*. Centre for Economic Performance, the London School of Economics and Political Science. Retrieved from https://cep.lse.ac.uk/pubs/download/dp1335.pdf

Greenleaf Enterprise. (n.d.). *Review of Scott's Technology Manual Assist Boning System: A Benefit Analysis*. Meat and Livestock Australia.

IFR Press Room. (2021). *Robot Race: The World's Top 10 automated countries.* https://ifr.org/ifr-press-releases/news/robot-race-the-worlds-top-10-automated-countries#:~:text=Japan%20(364%20robots%20per%2010%2C000,production%20are%20made%20in%20Nippon

International Federation of Robotics (IFR). (n.d.). *Robots benefit the US industry: 261,000 new jobs created in automotive sector alone.* https://ifr.org/ifr-press-releases/news/robots-benefit-the-us-industry-261000-new-jobs-createdin-automotive-sector

ISO. (2011). *Robots and robotic devices Safety requirements for industrial robots. Robots.*

Kowitt, B. (2018). *This startup is building the techiest indoor farm in the world.* http://fortune.com/2018/02/28/bowery-indoor- farm-technology/

Levert, C., & Héry, M. (n.d.). *Will technology improve health and safety at work?* https://www.greeneuropeanjournal.eu/will-technology-improve-health-and-safety-at-work/

Martech, M. (2013). *Positive Impact of Industrial Robots on Employment.* https://robohub.org/wp-content/uploads/2013/04/Metra_Martech_Study_on_robots_2013.pdf

Maximiliano, A. D., & Asha, B. (2019). *Which Countries and Industries Use the Most Robots?* https://www.stlouisfed.org/on-the-economy/2019/november/robots-affecting-local-labor-markets

McCarthy. (2020). *These are the countries with the highest density of robot workers.* https://www.weforum.org/agenda/2020/09/countries-comparison-robot-workers-robotics-change-tech-manufacturing

McKinsey Global Institute. (2021). *A Future That Works: Automation, Employment and Productivity.* https://www.mckinsey.com/~/media/mckinsey/featured%20insights/Digital%20Disruption/Harnessing%20automation

McKinsey Global Institute. (2021). *AI, Automation, and the future of work: ten things to solve for.* https://www.mckinsey.com/~/media/McKinsey/Featured%20Insights/Future%20of%20Organizations/AI%20automation%20and%

Murashov, V., Hearl, F., & Howard, J. (2015). Working Safely with Robot Workers: Recommendations for the New Workplace. *Journal of Occupation Environment Hyginic, 13*(3), D61–D71. doi:10.1080/15459624.2015.1116700 PMID:26554511

Nichols. (2020). *Business Benefits of Investing in Industrial Automation.* https://catalystforbusiness.com/business-benefits-of-investing-in-industrial-automation/

PwC. (2016). *Upskilling manufacturing: How technology is disrupting America's industrial labor force*. Price Waterhouse Coopers in conjunction with the Manufacturing Institute. http://www.themanufacturinginstitute.org/~/media/ E9F0B41DEC4F40B6AE4D74CBC794D26D/155680_2016_Manufacturing_ Labor_Force_Paper_final_(2).pdf

PwC Analysis. (n.d.). *How will automation impact jobs?* pwc.co.uk. https://www. pwc.co.uk/services/economics/insights/the-impact-of- automation-on-jobs.htm

Schroeder, N. (2021). *Global Manufacturing Labor Shortages*. globaledge.msu.edu. https://globaledge.msu.edu/blog/post/56844/global-manufacturing-labor-shortages

Scott Technology Ltd. (2019). *The Impact of Automation on Manufacturing*. Author.

Sepulveda, G. (2020). *Three reasons why businesses should invest in automation technology*. https://www.plantengineering.com/articles/three-reasons-why-businesses-should-invest-in-automation-technology/

Tortorella, G. L., Rossini, M., Costa, F., Portioli Staudacher, A., & Sawhney, R. (2021). A comparison on Industry 4.0 and Lean Production between manufacturers from emerging and developed economies. *Total Quality Management & Business Excellence, 32*(11-12), 1249–1270. doi:10.1080/14783363.2019.1696184

Uguina & Ruiz. (2019). Robotics and Health and Safety at Work. *International Journal of Swarm Evolution Computing, 8*, 176.

United Nations. (2017). *World Population Ageing 2017 - Highlights* (ST/ESA/ SER.A/397). Author.

Webster, C., & Ivanov, S. (2020). Robotics, artificial intelligence, and the evolving nature of work. In *Digital transformation in business and society* (pp. 127–143). Palgrave Macmillan. doi:10.1007/978-3-030-08277-2_8

Willcocks, L. (2020). Robo-Apocalypse cancelled? Reframing the automation and future of work debate. *Journal of Information Technology, 35*(4), 286–302. doi:10.1177/0268396220925830

Winick, E. (2021). Every study we could find on what automation will do to jobs, in one chart. *MIT Technology Review*. https://www.technologyreview.com/s/610005/ every-study-we-could-find-on-what-automation-will-do-to-jobs-in-one-chart/

Wright. (2021). *Automation and Its Impact on Employment*. https://www.nigelwright. com/uk/automation-and-its-impact-on-employment

Chapter 3

The Effect of Industrial Automation and Artificial Intelligence on Supply Chains With the Onset of COVID-19

Aditya Saxena
Amity University, Noida, India

Devansh Chauhan
Amity University, Noida, India

Shilpi Sharma
Amity University, Noida, India

ABSTRACT

This chapter discusses the various impacts of industrial automation and artificial intelligence in supply chains with the onset of COVID-19. The term industrial automation is influenced by rapid globalization and the various industrial revolutions that have caused the dire need for automation of industrial tasks to reduce human efforts. The chapter dives into the multiple fields affected by COVID-19 and how automation was used to deal with the situation, stabilize the supply chains, and maintain the profitability of organizations. Digital globalisation has led to the development of global supply chains. The use of technologies and cognitive automation and its effects have been discussed in the chapter. Machine learning has been used to get insight into the factors that affect supply chains and help their functioning.

DOI: 10.4018/978-1-6684-4991-2.ch003

INTRODUCTION

Supply chains and the way they function are always affected by events, and even the smallest of events may threaten their entire production line and profitability. To achieve stability and continuous profit regardless of how much a threat the future imposes creates an opening for the predictive applications of artificial intelligence to will be used. To understand the concept and implications of the topic, we first need to understand Digital Globalization and Industry 4.0 (Stentoft, J., & Rajkumar, 2020). We conducted a pilot search as part of the first phase to better our understanding of the examined field and the existing literature.

Digital globalization can be defined by the generation and exchange of data and information that primarily affects our ways of doing business and communication. We have done a complete systematic analysis of the situations and concluded about the exact impacts of covid to various supply chains and how it is different from the other pandemics that hit us earlier. This chapter also covers the various supply chains that were affected and what automation was used to maintain the profitability with the onset of covid (Schilirò, D, 2020).

BACKGROUND

Simple globalization can be termed as the interdependence of world economies.

Information, ideas, and innovations are being sent throughout the world via digital flows, widening involvement in the global economy and boosting digital globalization. The flow of data ensures that the digital era continues to grow and is forcing small businesses, freelancers, and even big government organizations to go digital and thus become more global and accessible to the people. Digital globalisations have led to the development of global supply chains. Take an example of a simple laptop; it is possible that the computer was designed in another place, parts were imported from another site, and used in another location. Global supply chains refer to using different areas for various supply chain tasks to reduce the cost price. Figure 1 is used to depict the multiple features of globalization.

Digitization and four the Industrial Revolution leading to Dependencies on AI:

Digitization paired with artificial intelligence is being used to automate industries and reduce human effort. The fourth industry evolution is deeply interconnected with the usage of artificial intelligence. Our entire ecosystem revolves around the objects of the internet of things(IoT), the need to automate tasks and the need for fast flow of data to enhance the working of these devices is dire. Various technical methods are used in digitization, such as predictive analysis, machine learning, extensive data analysis, and artificial intelligence are some of the many tools that help in automating and decision making.

Figure 1. Features of Globalisation

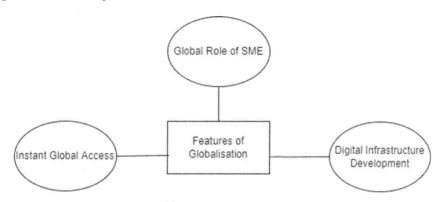

It is not wrong to say that Industry 4.0 will be of artificial intelligence, robotics, quantum technology, wearables, etc., and is bound to use the data generated and collected from the previous industrial revolution for this purpose. For the complete success of industry 4.0, several methodologies and strategies come. There are things that people need to change, which start from education.

People should not only have in-depth knowledge about their specializations but also should take into account the and at least have some knowledge of related fields.

It is being suggested that by 2030, 70 percent of the companies will be using at least one major artificial intelligence technology. People are concerned that artificial intelligence may lead to the loss of jobs. Skilled workers may not make a livelihood, contrary to the research conducted by Bughin et al., indicating AI can generate an additional $13 trillion in global economic activity by 2030, or a 16 percent increase in cumulative GDP over today. This equates to an annual increase in GDP of 1.2 percent (Fred Fontes, 2022).

Trade contributed to around 27 percent of the world's GDP in the 1970s, which rose to an astounding 60 percent in 2018, indicating the successful growth of globalization. But due to the widespread pandemic of covid -19, this shift is becoming more digital. The imposing lockdowns and instructing people to stay home would drain the current digital resources and increase a considerable demand for new resources. The amount of importance given to data and information can be understood because it is called the 'new oil' of the twenty-first century. More data also means that the quality of solutions that adopt artificial intelligence will increase as they have a much more readily available knowledge base. Countries like China, already knowing the importance of a digital economy, have already invested in digital technologies. The adoption of a digital economy has led to the room for people to participate and communicate digitally more and even better than before. Adopting a digital model directly depends on the type of sector you choose.

Figure 2. AI Technologies being Used in Industry 4.0

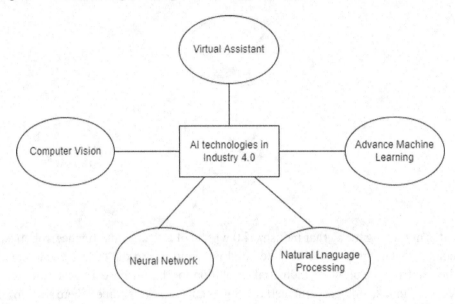

Adopting a digital model rather than a conventional one is no longer an option in the private sector. It is a compulsion regardless of the size of the private sector company; no matter how small, at least at one stage, it would need the help of the digital world. Many old and existing organizations are undergoing digital transformation, keeping their vision to enhance the customer experiences and take advantage of the existing technologies.

Tech giants like Facebook and Amazon have found ways to make a profit despite the challenges proposed by digital globalization and foreign expansion by making small start-ups and SMEs go global by using their platform and thus making them a primary player in the era of digital globalization. Undoubtedly, there are problems such as increased cyber-attacks and identity theft, but the advantages and features of digital globalization overshadow these flaws brilliantly.

These sudden implications have significantly affected our everyday lives and large-scale industries and organizations. Their manufacturing and working methods have changed drastically, which has led to significant changes in supply chains.

Many definitions of supply chains exist in the modern world, and the meaning of a supply chain is strictly environment-dependent. The complexity of the supply chain is also very much affected by the type and the scale of distance between the initial locations and where the final product has to be delivered. Supply chains show rapid fluctuations with the slightest change in abstract and non-abstract issues forced upon them.

An essential supply chain can be defined as a network between the suppliers trying to maximize their profit and deliver a final product delivered to the user. Supply chains involve a lot of co-dependency and inter-domain communication with the constraint of having significantly fewer or zero delays in transmission.

To a layman, communication may not be of the utmost importance. It is crucial for pharmaceutical-based supply chains as the temperature and delivery of drugs can be life or death for those in need.

The supply chain involves a lot of data being transmitted and processed, so a new problem arises: how we goanna process that data in time with taking most minor procedures. Here is when these nature-inspired optimization methods (e.g., ant colony optimization and genetic algorithms), game theory-based (e.g., cooperative models), market-based (e.g., negotiation and auction algorithms), decision theory-based (e.g., Bayesian approaches), and knowledge-based methods are examples of optimization methods.

Working of Supply Chains and Effects of Covid 19

There is a deep relation between machine learning and AI due to uncertainty in supply chains. We now have a much more uncertain environment than ever. We have a lot of covid induced problems like Lack of proper transport, raw material production, Lack of labour, and decreased consumption by users.

Figure 3. A Basic Supply Chain

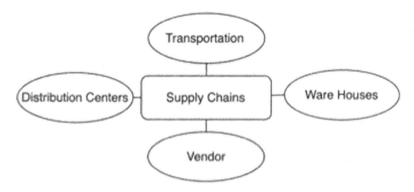

Due to covid and impositions of lockdowns in various regions. Taking the Indian example upon survey, many companies raised the issue about the difficulty in transport as raw material which needs to be transported being an essential aspect of the supply chain was not readily available due to government-imposed rules and regulations.

There are specific sectors where the covid 19 completely cut off the supply chain due to lack of raw material. Still, there are some sectors where the covid impact had a positive effect; instead, their revenue has gone up mostly where the manufacturing setup was producing essential products. If we talk about the indigenous supply chain which was severely hit due to lockdown and as far as global supply chain management they suffered around 70-90% reduction in terms of revenue, mainly the losses were seen in respect of air transports, vessels, trains and of course motor vehicle services. Thus the effect of covid on supply chains is highly dependent on the type of the industry and how the transportation services were affected by covid.

We already see a change from linear supply chains to more integrated networks linking numerous stakeholders due to the requirement for better visibility across hundreds or thousands of suppliers. Technologies such as IoT devices or sensors that offer valuable data on where items are in the chain and their condition — for example, products for which temperature monitoring may be crucial — enable this sea change (i.e., frozen foods, vaccines, or other medicines).

The epidemic has accelerated several pre-existing tendencies, including those in the supply chain: According to 64% of supply chain executives polled, the pandemic will hasten the digital transformation. The race for digital enablement and automation is on: The autonomous supply chain (e.g., warehouse and store robots, driverless forklifts and vehicles, delivery drones, and utterly automated planning) is either present by 2025, according to 52% of CEOs.

Covid introduced an atmosphere of uncertainty that hugely affects supply chains. To calculate these risks, artificial intelligence is used. Covid has also accelerated this digital transformation in many sectors, including education, and it has made it a compulsion for the world to learn technical skills. This can be seen with the massive increase in online education programs with the onset of covid.

Byju's would be the most appropriate example in this case which supports the argument that e-learning helps students learn faster as students can learn at their own pace.

Covid 19, contrary to other pandemics, has effects on both supply and demand.

And the global supply chains which may have shown resilience against natural disasters and pandemics in the recent decades are now compromised. Enhancing the strength of the supply chains is the main factor in reducing vulnerability and profitability in these critical times.

This is not the first time something significant has hit the global supply chains, tsunamis natural disasters keep posing a challenge to get all the time.

I consider the example of the tsunamis that hit japan in 2011 and 2004 tsunami in Indonesia we can see that the magnitude of the impact caused by covid is vital. If we look at the figure below closely, it is still highly uncertain when this will end. (Fred Fontes, 2022)

Covid 19 affected not only local supply chains but also global ones. The supporting pipelines of the GSC are affected by the covid-19, as shown in figure 4. Only those products which contribute the most are offered in the fixture and are classified.

This demonstrates how canceled flights, trade restrictions, and unavailable labor caused havoc in the global supply chains.

Generally, to understand the supply of products, we have classified the products into two categories, namely the functional products and general-purpose products. Functional products are those products that have both demand and supply under control and at normal levels.

Products that had minimal or less demand before the onset of covid were converted into innovative ones due to the sudden upsurge in demand and volatile needs of the customer. Products like face masks and face shields are the most appropriate examples of such products. The organization or people who understood this volatile supply and demand pattern were able to yield the maximum profit for their organization.

Figure 4. Transportation and Communication. (Schilirò, D, 2020)

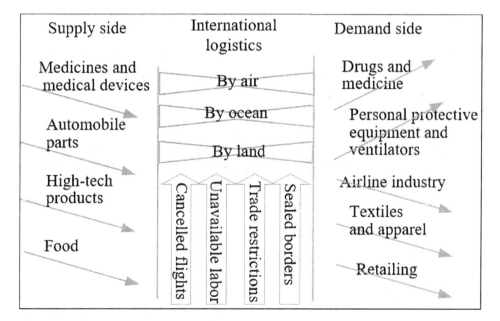

Commodities Having the most Disturbed Supply Chains

1. Tech Products: This is the most complex to understand gsc of this list. China is the caterer to the most tech demands of global consumers. The problem arose when lockdowns were imposed, and the demand for tech-related items

rose due to the rapid digitization of many businesses and organizations. This led to a massive uprise in the prices of even the most minor computers parts, a repercussion of the limited supplies. The impact was so disastrous on production that even multi-billion-dollar companies like Apple could not keep up with the production in China (F. Lauren, 2020). LG and Samsung also had to shut their productions in South Korea. Even the aeronautical sector's production was affected due to the imposed restrictions. Boeing and Airbus halted their productions. (F. Laureon, 2022).

2. Automobile Industry: The demand for automobiles saw a sudden downfall. In India, Nissan stopped production in the Chennai plant, and in countries like Egypt and the middle east, the production was halted. (S. Contractor, 2020)

3. Medicines: More than 40 percent of the world's supply of medicines directly or indirectly came from China. (M. Terry, 2018) Pharmaceuticals are facing shortages because many of them used to outsource their materials from China. The shortage of PPE kits led to the growth of many Indian Companies that started manufacturing PPE kits. These local small-sized startups have helped India to become the second-largest producer of PPE kits. These startups lowered India's gross import of the PPE kits a lot. (National Informatics Centre 's Website(NIC),2021)

4. Food Supply Chain Disruptions: The farming activities are affected by two main factors, i.e., Farming and transportation. The suspension of the rice exports from India due to labor. The gradual opening of lockdown and normalization of the world has led to the resuming of world food production. Still, some capital requiring factors such as fertilizers pricing labor shortages remain uncertain (The World Bank, Food Security and Covid's Website, 2020). The number of people facing chronic hunger saw an increase of 118 million in just one year e from 2019 to 2020. This indicates the discrepancy in the global food supply chains due to imposed lockdowns and trade restrictions, as of the current scenario of 2022

Distorted Demand of Supply Chains

1. Drugs and Medicine: There was a surge in demand for two main drugs due to the breakout of covid, namely chloroquine and hydroxychloroquine. In addition, regular medicines for fever, cough, cold, and body ache also increased. This increased demand made the government ration the number of items.

2. PPE and ventilators: We saw a shortage in the reserves of PPE kits and ventilators due to a covid outbreak. The dramatic rise in PPE demand required new production techniques to faster and cheaper produce the necessary product.

3. Airline Industry: Many operating authorities of various regions-imposed restrictions on air transport due to the spread of viruses. This caused many organizations to look for alternatives.
4. Textile and Apparel: This is one of the most labor-intensive industries, heavily affected by the lack of raw material and proper transportation facilities.

Plans for Dealing with the Situation

Supply Chains are susceptible when it comes to fault tolerance and disturbance. Even slight delays in the process can cause huge losses for firms. The timing of the first lockdown or restrictions was hugely important to people.

The term absolute response time (ART) indicates the number of days it took to impose restrictions after the virus hit.

Countries like the US had ART up to 62 days, making it easier for firms and suppliers to assess their risks.

Firms use these days to make and develop strategies and find alternatives to sustain their needs and profitability subtly. This requires tremendous skill and subtly studying the market, demand, and supply. For example, Tesla specifically started producing Ventilators, and fashion brands such as Louis Vuitton mainly started manufacturing masks for the consumers, contrary to popular belief. The production of sort of such products was a part of a large-scale mitigation plan. It required a lot of advertising and study of the market as these brands deviated from their original motive, which is significant. The most concrete example of this is BYD being the largest Bus manufacturer to the biggest producer of Face masks, which is pretty substantial. Table 1 clearly shows the increased automation use and what automation is being used to deal with the fluctuating demands and supplies.

In Figure 5, one can see the effect of globalization and changing tides in the Industries due to the usage of automation as the nature of jobs is changing in many fields and is going digital.

Implementation of AI

An AI is consistently implemented in supply chains for automation, keeping two factors as the primary goal:

1. The implemented intelligence should always choose the steps that would increase the total profit
2. The Automation used should at least have a basic sense of dealing with unseen situations.

Table 1. The Automation used to Manage the Supply Chains with the Onset of Covid 19

Industry / Field	Triggering Event	Automation Used/ Strategy Chosen	Effect
Mask and PPE kits	A sudden increase in Demand caused led to the rise in several start-ups that focussed on making masks and PPE kits.	SWACHH machine developed by Saral Design Solution Producing SITRA certified masks(Raju et al., 2022) Development of Nanofiber based mask	I am meeting the Huge Demand and Reduction in the pricing of the N95 Mask.
Education and Academics	Imposed Lockdowns caused schools' to shut down, leading to direct effects on more than 1.2 billion children. The net worth of Online education is projected to reach 450 billion Dollars by 2025(World economic Forum, The COVID-19 pandemic has changed education forever, 2021).	Development of Virtual Labs and Online software for educational purposes.	A strain on the supply chains of tech companies with the sudden uprise in demand for laptops and Tablets.
Health Sector	An increase in demand for specific drugs and PPE kits caused a lot of problems in the healthcare supply chains as 80% of the world imported masks from China in the pre covid era (Vaishnavi Dayalani, 2021)	Redundancy building in supply chains was implemented using artificial intelligence for distribution monitoring (Vaishnavi Dayalani, 2021). Diversification of the distribution system was done.	Even though initially many countries failed, more or less the demands of the people for medical needs were met.
Tech Companies and Call Centres	The shutting down of offices and call centers was the primary triggering event for the increased rapid digitization with the onset of covid. According to the sources in (Top COVID Challenges for Contact Centers in 2021.CGS website,2021)	Implementation of work from home strategies and development of work from home platforms. This trend can be seen in the given table. More focus was given to the Loom strategy.	The physical dimension of work was elevated for the first time after the onset of covid(Susan Lund et al, 2021). Change like work has caused many people to change their occupations. According to (Susan Lund. et al., 2021), more than 100 million workers or 1 in 16 will have to find a new career. This trend can be seen in figure 5(Susan Lund et al., 2021).
Automotive Industry	There was a prolonged demand truncation due to the imposed lockdowns and restrictions.	Revision of forecast instructions and prioritizing downsize given the market trends and	The companies were able to sustain the entire period of deteriorating demand.

Figure 5. The Shifting Tides of Employment (Susan Lund et al.,2021)

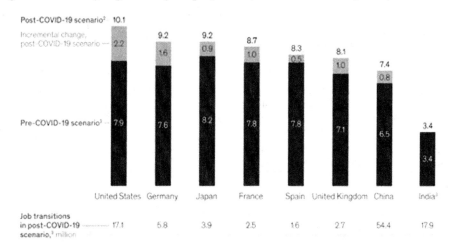

AI and Big Data also supplement each other to provide generally more results in a significant way. First, information is fed into the AI engine, which subtly improves the AI's intelligence. Next, much less human interplay is required for the AI to real characteristically correct. Finally, fewer human beings are needed to run AI. Specifically; the nearer society may be to satisfy the overall ability of the persevering AI cycle.

In addition to data from industrial information systems, the Internet of Things (IoT) enables distant sensors to connect with significant networks or even other goods, allowing for the collection of more relevant data from manufacturing shop floors. Radiofrequency identification (RFID), wireless sensor networks (WSN), and Bluetooth low energy devices, such as beacons, are among the technologies employed (Xu, Xu, and Li, 2018; Cavalcante et al., 2019).

Many recent studies on training machine learning and neural network models for business forecasting have shown that AI can perform a wide range of intelligent business activities. Various forecasting approaches, for example, are introduced for multiple objectives, such as recognizing credit risk (Zhu et al., 2019), minimizing the bullwhip effect (Singh and Challa, 2016), inventory level (Paul, Azeem, and Ghosh, 2015), and consumer demand (Paul, Azeem, and Ghosh, 2015). (Kochak and Sharma, 2015). (Cavalcante et al., 2019) concentrated on the use of supplier selection by creating supplier risk profiles. (Lyutov, Uygun, and Hütt, 2019), On the other hand, focused on client management. (Goli et al., 2019) and (Baryannis, Dani, et al., 2019) showed how to forecast and manage risks in product portfolios and the supply chain.

Table 2. Advantages of Using Artificial Intelligence in Supply Chains

Accurate Inventory Management	Managing inventory-related variables like processing, picking, and packing is challenging. Using artificial Intelligence management is improved, and demand is forecasted. AI is also used for discovering patterns.
Warehouse Efficiency	AI can ensure timely item retrieval and ensure a smooth and timely product flow directly to the consumer.
Enhanced Safety	Automation is being used to enhance the security of production lines. Ensuring the quality and various standard production parameters increases the overall safety of the workers working.
Decreasing Operating Costs	The use of Reinforcement learning-based agents increases the amount of work being done simultaneously compared to Humans.

Expert Systems - Expert system is a term used to mainly represent computer programs capable of making decisions like Humans and understanding the scenario and situation. Expert systems are composed of these four pretty essential components, i.e., or so they specifically thought. Knowledge base, justifier, Interference engine, and user interface are pretty significant. According to reports, using an expert system has increased IBM's product output by 35% and saved approximately $10 million of notably capital expenditure in a colossal way.

The ability of AI to specifically underlie the decision-making process in supply chains generally has sparked a lot of thought on how commonly AI may be used to, for all intents and purposes, improve the supply chain's generally long-term performance and competitive advantage, i.e.,, innovation. While previous research has suggested that process innovation enhances SCRes in uncertainty, leading to higher SCP. Developing AI-driven innovation can speed up the decision-making process in discovering, creating, and testing innovative solutions contrary to popular belief. Contrary to popular belief, researchers call this decision-making process, specifically at the foundation of innovation design. Long considered a vital aspect in SCRes development and SCP improvement, the impact of AI-driven innovation

Table 3. Challenges of Artificial Intelligence in Supply Chain

System Complexity	AI systems are relatively complex to set up, and many industry experts may be needed to monitor A's work. I based systems
Scalability Factor	The generic AI-based solutions need to be scaled for different needs.
Cost Edured for Training	AI requires a good amount of upfront capital investment and may create a training phase that hinders the initial production.
Operational Costs Involved	Machines have an extensive network of processors, storage, and networking devices working together to achieve the goal and require constant maintenance and support from technicians and industry experts.

on constructing resilient supply chains through information exchange, information processing, and system integration significantly.

Technological Infrastructure

The technological infrastructure for collecting the data centrally in real-time and with data from the previous observation can be used in training the AI model, which is quite significant. Many industrial information systems (IoT devices) particularly have integrated support for this communication between the information provided by the hardware and the model like MES (manufacturing execution system), which is mostly quite significant. Hardware like MES, SCADA (Supervisory control and data acquisition), PLC (Programmable logic controller system), for the most part, uses the real-time dynamics and the controls of the actual entire supply chain or the supply chain management system majorly.

The Data Collection Infrastructure was created to centralize the collection of real-time and historical data for model training. Real-time and dynamic control and monitoring of the entire process of the production system is done by methods such as the manufacturing execution system (MES). There are other systems like supervisory control and data acquisition systems and the programmable logic controller system, which is responsible for directly controlling the reaction parameters of machines. Warehouse management systems(WMS) are also used to support informatory communication and data.

The second component of the technological infrastructure is the model training phase which consists of cloud computing, big data analytics, and machine learning. In the first component, they are linked to the Data Collection Infrastructure to allow decentralized data and secure big data analysis models to work on the acquired data.

The researchers have used the logistic regression classifier pretty well due to the explicit facts. Logistic regression fashions mostly are a beneficial device for studying binary and particular points due to the fact they can kind of help you behavior a contextual evaluation to apprehend the relationships among variables, mostly take a look at for differences, estimate impacts, actually make predictions, and actually put together for destiny situations, or so they thought.

This strategy, however, is complicated to implement. It necessitates the existence of sufficient CPU memory to store the information acquired by training datasets and a high degree of IT architectural requirements in terms of security, privacy, and resource restrictions (Choi, Wallace, and Wang 2018).

Many of the AI algorithms tested within the research paper promote creativity to offer solutions that motivate advanced SCP. Even though the tremendous capacity of AI approaches to growth SCRes, there may be a lack of hobby within the modern research on AI-driven innovation. Influences SCRes, and to what diploma deliver chain dynamism (SCD) and deliver chain cooperation (SCC) affect this connection.

Information Processing Capabilities of Artificial Intelligence (AI)

Artificial Intelligence (AI) is a system's ability to kind of gather learnings with the aid of using assessing enter from the outside surroundings after which using the one's understandings to evolve or create new plans in reaction to the one's chances in a massive way. This class accommodates techniques and algorithms that permit us to study data, whether or not we are no longer aware of the very last output forms significantly. The field of artificial intelligence is not new in and of itself. However, since its inception in the 1950s, it has seen significant growth and catastrophic decline phases. The development of big data and the greater use of AI in operations, manufacturing, and supply chain management have contributed to a renewed focus on this computing technology.

Artificial Intelligence-Driven Supply Chain Innovation (SCI)

SCI is described as a gradual change in supply chain technology, process, or network that may, for the most part, be pretty leveraged to increase new value generation for stakeholders, which for the most part is quite significant.

SCI comprises all measures targeted at mitigating environmental uncertainty through information processing and technical innovation to solve supply chain challenges and identify new ways to improve operations, demonstrating how SCI comprises all measures targeted at mitigating environmental uncertainty through information processing and technical innovation solve supply chain challenges and identify new ways to improve operations particularly, or so they essentially thought. Supply chain innovation, in particular, relies heavily on for all intents and purposes current technology and processes, as well as substantial improvements in product, service, or strategy that improve efficiency and value delivery to the end-user in a subtle way. Corporations, for example, are increasingly using AI techniques to overcome the information processing constraints imposed by SCI, resulting in remarkably fresh ways to product creation and supply chain management, demonstrating that supply chain innovation, in particular, relies heavily on basically current technology and processes, as well as substantial improvements in product, service, or strategy that kind of improve efficiency and value delivery to the end-user, or so they mostly thought.

Supply Chain Resilience (scres)

Supply chain resilience (SCRes) is explicitly at the pinnacle of the C-interest suite these days, definitely due to the COVID-19 epidemic. SCRes has involved approximately the delivery chain's capability to react to unexpected disruptions and, as a result, attain former or maybe basically advanced overall performance levels, which is pretty significant. SCRes is monitored at some stage in all three phases of a disruption: proactive, pre-disruption preparedness, reaction, and recovery (reactive, post-disruption), which for the most part is quite significant. The cause of this definition can, for the most part, be determined in preceding research, which for the most part, highlights the idea of regaining equilibrium following a disruptive event.

OIPT (Organizational Information Processing Theory)

The development of organizational skills to meet their information processing requirements is addressed by OIPT. According to OIPT, businesses must handle information with rising uncertainty to maintain a given degree of performance. An essential organizational competency is the ability to process data in the face of risk, volatility, and dynamism. In this study, AI is viewed as an information-processing tool that should be built-in to reduce operational difficulties and uncertainties. Furthermore, the OIPT recommends that businesses establish capacity builders and information processing capabilities to deal with supply chain interruptions.

International logistics disruption shut downtown of commercial and, in many cases, passenger aviation had a severe effect on air cargo.

Table 4. Benefits of Using AI in Supply Chains

Planning and Scheduling	A global supply chain consists of global networking, which requires an end-to-end solution for managing various parts like network, business, transport, which is pretty significant. As the number of parameters increases, the visibility tends to decrease. Thus, mainly AI enhances visibility by extending its use cases in logistics; managers can predict the Bottlenecks and abnormalities.
Analytical Insights	Analytical insights are given and implemented in supply chains using cognitive automation. Further explained in table xx.
Operational Efficiency Increased	The data generated by millions of IoT devices requires a lot of focus, computation power, and work. Automating these tasks using AI is a massive boost in Operational efficiency.
Streamlining Enterprising Resource Planning	Implementing AI in supply chains helps streamline the ERP framework and efficiently connect the various components.

Cognitive Automation, A Solution for Supply Chain Management

Cognitive automation is being used in many supply chains to make supply chains more resilient and increase the process's overall stability. It has emerged as the leading form of automation used to enhance various fields. Surveys show that 31%

of the businesses have at least fully automated one of their organization's work functions. Adopting Artificial Intelligence is a smart strategy as it has shown an increase in profit by 28%. Cognitive automation is the collective use of AI, machine learning, and human intelligence to respond to any scenario more prepared. Supply chains possessing the correct data at the right time can be described as self-healing supply chains.

A self-healing supply chain saves time for data scientists and engineers who no longer have to retrain a model or an AI. The point that makes a huge difference is the use of experience replay and studying the effects of decisions that were made in the past.

Consider a large-scale multinational petrochemical company that aims to provide sustainable products as the demand for chemicals is bound to increase. The company used cognitive automation that related the order details from the beginning and the details of third-party vendors. With this amount of end-to-end visibility, the company could make in-time deliveries, and the execution time for operations was reduced.

The fields where cognitive automation is used in supply chains are as follows.

1. Labour Scheduling: People are critical resources of any functioning Supply Chain. Flagging of the impending problems can be done using cognitive automation technologies.
2. Capacity Planning: It is challenging to predict the amount of transportation and storage one might need as the demand for most of the products fluctuates and needs constant monitoring. Cognitive automation can be used by predicting variability and deducing real-time options that maximize capacity planning.
3. Container Space Optimization: Surveys have shown that the visibility of supply chains is abysmal, and ensuring transparency for full utilization of space and capacity is lacking in many modern-day supply chains. Only 6 percent of the significant supply chains have full visibility and complete updates accurate up to minutes and seconds (Fred Fontes, 2022). Who can use cognitive automation to correlate data with the container details and ensure that every container leaving the dock is utilized to the fullest extent?
4. Resolving the issue of cross-docking and Direct shipment: Increased globalization has made docking dormant, and direct shipment is the fastest way. Identification of bottlenecks and enhancement of shipping networks can be made via cognitive automation.
5. Disaster Recovery: The list of factors that can affect a supply chain in predictable and unpredictable ways is never-ending. There must always be a backup plan for situations like natural disasters, political—ethnicities, lack of workers, etc. Who can use cognitive automation to recover from disasters and provide a suitable alternative to any situation?

MODEL

Purpose of the Proposed Model

The proposed machine learning model is used to determine the late delivery risk and optimize the supply chain by early detection of the delivery delays and can help the organizations to make decisions about the item that is causing the delay and optimize the supply chain accordingly in a subtle way. Extensive machine learning algorithms were used for this purpose with appropriate feature selection, and the model is chosen by understanding the type of data we are working with.

The Dataset used for the Model dataset of the proposed model has been taken from Kaggle and has approximately 90000 thousand data entries taken over three years. The dataset has 54 features; analysis and data preprocessing are done to get the desired data with limited features to reduce the computing power, directly fed into the machine learning model. The main features that are selected for this purpose are as follows. The demand for any product is mainly catered by proper and timely supply. For the store to be on time, it is essential to establish an appropriate system that ensures the delivery of the raw product at a suitable place. There are a lot of features in the given dataset, out of which a few a selected in the step of data preprocessing and cleaning.

1. Type
2. Days for shipping (Real)
3. Shipping Mode
4. Late delivery Risk

ML Classifier Used (Christina et al., 2020)

The researchers have used the logistic regression classifier due to the explicit facts in an actual big way. Logistic regression fashions are a beneficial device for studying binary and generally specific points due to the fact they can help you behavior a sort of contextual evaluation to apprehend the relationships among variables, mostly take a look at for differences, estimate impacts, specifically make predictions, and precisely put together for destiny situations, which generally is quite significant.

Consider the simple linear regression model, in which Y is a continuous variable and X is a binary variable. When $X = 0$, $E(Y|X=0) = \beta_0$ and when $X = 1$, $E(Y|X=1) = \beta_0 + \beta_1$. i.e. the mean difference between when $X = 0$ (reference group) and when $X = 1$. (Comparison group).

Consider the simple linear regression model, in which Y is a continuous variable and X is a binary variable. $E(Y|X=0) = 0$ when $X = 0$ and $E(Y|X=1) = 0 + 1$ when

Figure 6. Value for odds

$$\log(odds) = \log(\Omega) =$$

$$\log\left(\frac{E(Y|X)}{1 - E(Y|X)}\right) = \beta_0 + \beta_1 X$$

$X = 1$. As a result, we take 1 to signify the mean difference between the two groups, i.e., the mean difference between when $X = 0$ (reference group) and when $X = 1$. (comparison group).

Consider the following basic example in which Y and X are binary variables with values of 0 or 1. Because we are no longer looking at means when using logistic regression, the meaning of 1 changes. Remember that logistic regression has a model $\log(E(Y|X)/(1-E(Y|X))) = \beta_0 + \beta_1 X$ or for simplification's sake, $\log(\pi/(1-\pi)) = \beta_0 + \beta_1 X$.

Figure 7. Value for different Binary variables

X = 0	$\log\left(\frac{\pi_0}{1 - \pi_0}\right) = \beta_0$
X = 1	$\log\left(\frac{\pi_1}{1 - \pi_1}\right) = \beta_0 + \beta_1$

If we want to interpret β_1 from the two examples above, we'll approach it as if it were a straightforward linear regression. That is, β_1 is obtained by subtracting the result when $X = 1$ from when $X = 0$.

Figure 8. Value for β_1

$$\beta_1 =$$

$$\log\left(\frac{\pi_1}{1 - \pi_1}\right) - \log\left(\frac{\pi_0}{1 - \pi_0}\right) =$$

$$\log\left(\frac{\pi_1/(1 - \pi_1)}{\pi_0/(1 - \pi_0)}\right)$$

Libraries Used:

Pandas

Numpy

Scikit

matplot-lib

Seaborn

Parameters used:

Random_state:

random_stateint, RandomState instance, default=None

Used when solver == 'sag,' 'saga,' or 'liblinear' to shuffle the data.

Max_iter:

max_iterint, default=100

The maximum number of iterations taken for the solvers to converge.

Feature Selection:

The data has 180519 rows × 46 columns; seeing the amount of data and reducing the computing power, the researchers decided to run feature selection (feature selection is a process of selecting features from a given model that relates to the model researchers make while also reducing the computing power), from the heatmap. It is visible that only five elements are related to the value we are predicting.

Figure 9. Co-Relation Matrix of all features

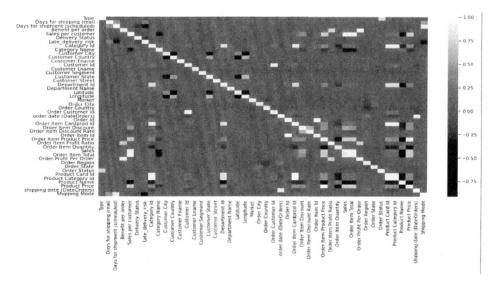

From the 46 features, we come down to 5 now. Again, the model and the correlation matrix deduced that two columns/elements in the training dataset have high co-relational value, ruining the model accuracy.

Figure 10. Co-relation Matrix feature Selection using Confusion Matrix

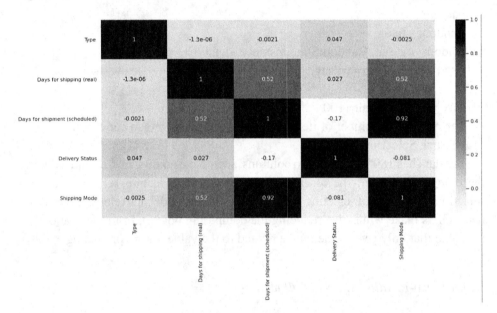

The above figure shows that the two features, 'Days for shipment' and the feature 'Days for Shipment,' are highly correlated. Having highly correlated features in the training model will reduce model accuracy, so we remove one of them.

After removing one of them, we have finally concluded the feature selection procedure.

It can be concluded that the model is working perfectly fine with an accuracy of 97 percent, and it is essential to make feature selection because of the high computing power that it requires.

The following graph shows the feature importance.'

RESULT

The logistic regression model has an accuracy of 97.71 percent, which means that the model can accurately predict the late delivery risk by 97.7 percent.

Figure 11. Co-relation Matrix for the final model

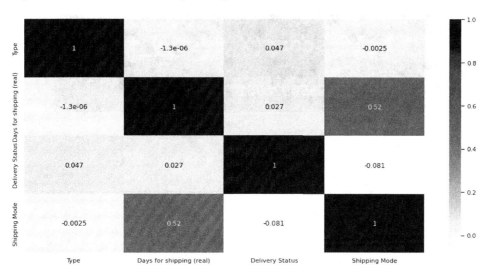

Figure 12. Feature importance

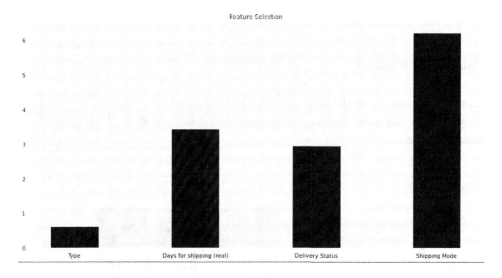

The above flow-chart shows the researchers' steps in order to make the desired model.

Figure 13. The final accuracy of the model

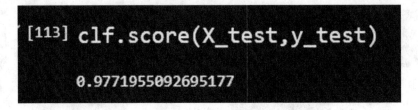

Figure 14. Flowchart

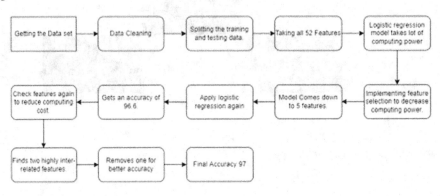

Data Visualization

It can be concluded that the model is working perfectly fine with an accuracy of 97 percent, and it is essential to make feature selection because of the high computing power that it requires.

Figure 15. Confusion Matrix of the model

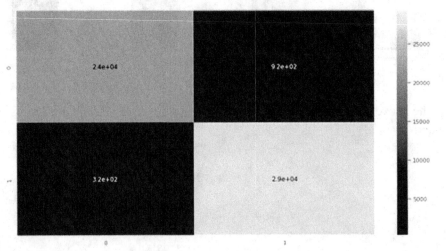

From the confusion, we can see that the model is perfect, with an accuracy of 97.7. The number of false positives and false negatives was very low.

The following graph shows the visualization between the predicted and the actual values of the model.

Figure 16. Visualization of the model's accuracy

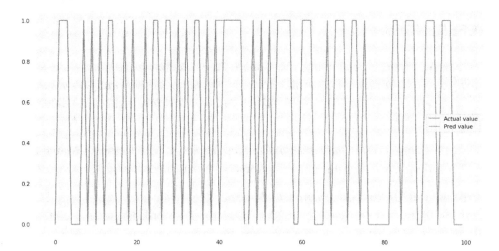

FUTURE RESEARCH DIRECTIONS

This paper focuses on how artificial intelligence can be integrated with the supply chain management system to create the most optimized, error-free, and profitable supply chain. Integration of Artificial Intelligence in supply would further help the world combat situations like covid-19 to maintain the collection of everything, most notably the essential items, and help to various challenges faced in the supply chain like difficult demand forecasting, port congestion by predicting deliveries, material scarcity by forecasting demand using consumer data.

CONCLUSION

The chapter gives a detailed insight into the working of supply chains. It provides a detailed overview of the use of artificial intelligence on supply chains to manage various processes and enhance the working of the supply chains. The chapter also discusses the impact of covid 19 on the different industries and supply chains and how the entire situation of covid was dealt with; the increased globalization and the ongoing digitation also play a crucial role in modernizing and developing supply chains. The imposed restrictions and lockdowns boosted the process of digitation and industrial automation, which are directly related to supply chains. A simple machine learning model has been made, which indicates late delivery of products and can be used by organizations to focus on the products that are being delayed and make suitable changes to their supply chains. The real impact of covid on various sectors has been written into simple crux points for simple understanding.

REFERENCES

Belhadi, A., Mani, V., Kamble, S. S., Kah, S. A. R., & Verma, S. (n.d.). *Artificial intelligence-driven innovation for enhancing supply chain resilience and performance under the effect of supply chain dynamism: an empirical investigation.* Academic Press.

CGS. (n.d.). *Top COVID Challenges for Contact Centers in 2021.* Retrieved March 12, 2022, from https://www.cgsinc.com/blog/top-covid-challenges-contact-centers-2021

Christina. (2020). *An Introduction to Logistic Regression for Categorical Data Analysis.* Retrieved March 12, 2022, from https://towardsdatascience.com/an-introduction-to-logistic-regression-for-categorical-data-analysis-7cabc551546c

Cognitive Automation Community. (2020). *5 Ways Cognitive Automation is Transforming Supply Chain Processes.* Retrieved March 5, 2022, from https://www.cognitiveautomation.com/resources/5-ways-cognitive-automation-is-transforming-supply-chain-processes

Contractor, S. (2020). *Coronavirus pandemic: Nissan temporarily halts operations in Asia, Africa, Middle East.* Carandbike. Available: https://auto.ndtv.com/news/coronavirus-pandemic-nissan-temporarily-haltsoperations-in-asia-africa-middle-east-2200191

Dayalani, V. (2021). *12 Charts That Show The Rise Of Indian Tech During Covid.* Retrieved October 13, 2001, from https://inc42.com/datalab/12-charts-that-show-the-rise-of-indian-tech-after-lockdown/

Deloitte. (2022). *Understanding COVID-19's impact on the automotive sector.* Retrieved April 25, 2022, from https://www2.deloitte.com/us/en/pages/about-deloitte/articles/covid-19/covid-19-impact-on-automotive-sector.html

Dubey, R., Gunasekaran, A., Childe, S. J., Bryde, D. J., Giannakis, M., Foropon, C., Roubaud, D., & Hazen, B. T. (2020). Big data analytics and artificial intelligence pathway to operational performance under entrepreneurial orientation and environmental dynamism: A study of manufacturing organizations. *International Journal of Production Economics*, *226*, 107599. doi:10.1016/j.ijpe.2019.107599

Fontes, F. (2022). *Cognitive Automation Is Building Self-Healing Supply Chains*. Retrieved April 26, 2022, from https://www.supplychainbrain.com/blogs/1-think-tank/post/34424-cognitive-automation-is-building-self-healing-supply-chains

Grover, P., Kar, A. K., & Dwivedi, Y. K. (2020). Understanding artificial intelligence adoption in operations management: Insights from the review of academic literature and social media discussions. *Annals of Operations Research*. Advance online publication. doi:10.100710479-020-03683-9

Lauren, F. (2020). *iPhone manufacturing in China is in limbo amid a coronavirus outbreak*. Available: https://www.cnbc.com/2020/02/10/coronavirus-leaves-status-of-apple-manufacturing-in-china-uncertain.html

Lund, S., Madgavkar, A., Manyika, J., Smit, S., Ellingrud, K., & Robinson, O. (2020). *The future of work after COVID-19*. Retrieved March 12, 2022, from https://www.mckinsey.com/featured-insights/future-of-work/the-future-of-work-after-covid-19

Mahmoodi, F., Blutinger, E., Echazú, L., & Nocetti, D. (2021). *COVID-19 and the health care supply chain: impacts and lessons learned*. Retrieved April 22, from https://www.supplychainquarterly.com/articles/4417-covid-19-and-the-health-care-supply-chain-impacts-and-lessons-learned

National Informatics Centre(NIC), Government of India. (n.d.). *Technology based startups played a crucial role in converting India from importer to second largest manufacturer of PPEs*. Retrieved 31 December, 2020, from https://dst.gov.in/technology-based-startups-played-crucial-role-converting-india-importer-second-largest-manufacturer

Raju, G. S., Kumar, N. S., & Nikkat, S. (2020, December). Technology based startups pivoting for sustainability: Case study of startups. *IOP Conference Series. Materials Science and Engineering*, *981*(2), 022083. doi:10.1088/1757-899X/981/2/022083

Schilirò, D. (2020). *Towards digital globalization and the covid-19 challenge*. Academic Press.

Shukla, A. (2020). *Coronavirus fears send the aerospace industry into a tailspin*. Available: https://www.rediff.com/business/report/covid-19-fears-send-aerospace-industry-into-tailspin/20200322.htm

Stentoft, J., & Rajkumar, C. (2020). The relevance of Industry 4.0 and its relationship with moving to manufacture out, back, and staying at home. *International Journal of Production Research*, *58*(10), 2953–2973. doi:10.1080/00207543.2019.1660823

Terry, M. (2018). *Recent drug scandals in China spotlight potential global supply chain issues*. Available: https://www.biospace.com/article/ recent-drug-scandals-in-china-spotlight-potential-global-supply-chain-issues/

The World Bank. (n.d.). *Food Security and Covid*. Retrieved January 20th, 2022, from https://www.worldbank.org/en/topic/agriculture/brief/food-security-update

Wamba, S. F., Dubey, R., Gunasekaran, A., & Akter, S. (2020). The performance effects of big data analytics and supply chain ambidexterity: The moderating effect of environmental dynamism. *International Journal of Production Economics*, *222*, 107498. doi:10.1016/j.ijpe.2019.09.019

World Economic Forum. (2020). *The COVID-19 pandemic has changed education forever. This is how*. Retrieved February 12, 2022, from https://www.weforum.org/agenda/2020/04/coronavirus-education-global-covid19-online-digital-learning/

Xu, Z., Elomri, A., Kerbache, L., & El Omri, A. (2020). Impacts of COVID-19 on global supply chains: Facts and perspectives. *IEEE Engineering Management Review*, *48*(3), 153–166. doi:10.1109/EMR.2020.3018420

KEY TERMS AND DEFINITIONS

Artificial Intelligence: The intelligence shown by machines to give logical answers and make smart decisions to questions utilizing a knowledge base.

Cognitive Automation: It is the collective use of AI, machine learning, and human intelligence to respond to any scenario more prepared.

COVID-19: An infectious disease caused by SARS-CoV-2 virus.

Digitization: The process of conversion of various business and industries going global.

Globalisation: It refers to the process of interconnecting the various organisations on a global level.

Industry 4.0: The industrial revolution which leaded after industry 3.0 and utilizes the latest global technologies such as artificial intelligence and machine learning.

Machine Learning: The branch of computer science that is related to training models on data set and uses advance mathematics and algorithms for prediction.

Supply Chain: It refers to the entire process of supplying a product to the consumer with all the technicalities involved.

Chapter 4
Convolutional Neural Networks and Deep Learning Techniques for Glass Surface Defect Inspection

Eduardo José Villegas-Jaramillo
Universidad Nacional de Colombia - Sede Manizales, Colombia

Mauricio Orozco-Alzate
Universidad Nacional de Colombia - Sede Manizales, Colombia

ABSTRACT

Convolutional neural networks and their variants have revolutionized the field of image processing, allowing to find solutions to various types of problems in automatic visual inspection, such as, for instance, the detection and classification of surface defects in different types of industrial applications. In this chapter, a comparative study of different deep learning models aimed at solving the problem of classifying defects in images from a publicly available glass surface dataset is presented. Ten experiments were designed that allowed testing with several variants of the dataset, convolutional neural network architectures, residual learning-based networks, transfer learning, data augmentation, and (hyper)parameter tuning. The results show that the problem is difficult to solve due to both the nature of the defects and the ambiguity of the original class labels. All the experiments were analyzed in terms of different metrics for the sake of a better illustration and understanding of the compared alternatives.

DOI: 10.4018/978-1-6684-4991-2.ch004

INTRODUCTION

Seeking for defects in surfaces such as glass sheets is a task that can be performed from different approaches, among which deep learning (DL) stands out as an alternative that offers good results when applied in its different variants such as convolutional neural networks (CNN), deep networks and residual learning (RL)-based networks (Zhou et al., 2019), which can be complemented by using techniques such as data augmentation (DA) (Shorten and Khoshgoftaar, 2019) and transfer learning (TL) (Hafemann et al., 2015). The number of model configuration alternatives might become a combinatorial problem, making difficult for the practitioner to design a system that fulfills both performance requirements and practical constraints. Therefore, the objective of this chapter is to present a comparative study of different models that, based on DL, address the problem of classifying thirteen categories of defects that appear in images from a public glass inspection (GI) surface. The GI dataset seems to be difficult to classify due to its heterogeneity and the potential ambiguities in the original labeling; moreover, to the best of the authors' knowledge, the GI dataset has not been used before in other studies since the authors did not find citations to it.

The comparative study consists in ten experiments with several variants of the dataset, CNN architectures, RL-based networks, TL, DA, and (hyper)parameter tuning including combinations of various configuration alternatives. The evaluation was made according to global metrics such as accuracy (reporting its minimum, maximum and average values) and the micro and average measures of precision, recall and F1-score; other more specific performance measures were also considered, including the confusion matrix and the classification report by class, which were complemented with learning-validation curves and predictive classification graphs.

Regarding the design of the classification system, two neural network architectures were adopted, namely: i) a CNN and ii) a RL that makes use of TL by incorporating a previously trained Inception ResNet (Deng et al., 2009). The chosen networks were combined with DA techniques that were applied either in a previous phase or at training time. The best solution was obtained with a model using RL-based networks and DA, reporting an average accuracy of 0.8365, which the authors have considered good given the challenging nature of the problem.

The remaining part of the chapter is organized as follows. Essential concepts of visual inspection techniques, DL, neural networks, RL-based network, TL, and DA are described in section Background and related work. The subsequent section presents the Method for the comparative study, describing the different variants of the datasets, the CNN models, and the configuration of the experiments. Afterwards, the Experimental results and discussions achieved are presented and finally, the Conclusion is given.

BACKGROUND AND RELATED WORK

Methods for Automatic Visual Inspection

The use of techniques for automatic visual inspection in manufacturing processes and especially for seeking defects in surfaces is becoming more frequent every day. To solve this problem, several strategies are used, including projection methods (Chen and Perng, 2011), filter-based approaches (Lin and Ho, 2007) and standard machine learning approaches (Chen and Hsu, 2007). Most of these strategies assume certainty in the class labels assigned by experts to the training examples; however, label weakness or ambiguity (Zhou, 2017) may appear in some datasets requiring the application of non-standard approaches (Mera et al., 2016).

Other approaches use neural networks and their variants. In (Weimer et al., 2015), for example, the authors use a deep convolutional network (DCNN) and classify twelve categories of defects from the DAGM2007 dataset; furthermore, they analyze the impact of applying different strategies used in extracting patches from images to train a neural network that, by using techniques such as DA, solves the problem of class imbalance. It is also important to highlight combinations of techniques such as those used in (Medina et al., 2011) for the detection of defects in steel sheets. In that study, the authors initially extract features based on shape and brightness, and then train three different classification techniques to create an ensemble, namely: an artificial neural network, a *k*-nearest neighbors algorithm, and a naive Bayes classifier.

Deep Learning and Convolutional Neural Networks

DL allows computational models, based on multiple processing layers, to learn and represent data with a large number of abstraction levels (Hatcher and Yu, 2018). DL includes CNN, deep belief networks, deep Boltzmann machines, RL-based networks, TL, and encoders; all these models and techniques are particularly relevant for solving problems related to computer vision, face recognition, object detection, autonomous car driving, and visual inspection systems (Voulodimos et al., 2018). DL has shown good potential due to its capabilities for dealing with large and complex datasets; furthermore, unlike traditional architectures, DL can be applied to all types of data (audio, numeric, text or any other combination). On the other hand, unsupervised DL has presented good results for clustering and statistical analysis and is frequently applied for computer vision (Hatcher and Yu, 2018).

The most popular DL method is CNN: an advancement of the well-known neural networks. CNNs are based on multilayer models that use a single neural network composed by combinations of convolutional layers, maximum clustering layers, and fully connected layers, which exploit spatial correlation by applying local connectivity patterns between neurons in adjacent layers to identify relationships between objects (Lazebnik, 2015, Smith et al., 2021). Its application in surface defect inspection can be developed either in a single stage or in several ones. An example of the first case is presented in (Cha et al., 2017), where the CNNs carry out both feature extraction and classification, including a detection method for five types of damages: concrete crack, steel corrosion with two levels (medium and high), bolt corrosion and steel delamination. An example of the second case is presented in (Yu et al., 2017), where a two-stage DL-based method is used to look for surface defects in industrial environments, combining a segmentation stage and a detection stage, which consist of two separate fully convolutional networks; additionally, the system is trained with patches cropped from the images in order to learn from a limited amount of data. Similarly, in (Tabernik et al., 2019), a DL model based on two CNNs is proposed: the first one aims to segment the images and detect small defects on the surface, while the second receives the output generated in the segmentation and performs classification. The experiments in the latter work are carried out to inspect the quality of electrical commutators and demonstrate that the DL model can be successfully used even when only a few defective samples are available for training.

Residual Learning-based Networks

RL-based networks are a family of extremely deep architectures that use a residual unit in which signals can propagate directly —backward or forward— from one block to any other, thus allowing to bypass connections and their subsequent activation. In (He et al., 2016), theoretical and practical evidence is provided regarding the advantages of using a RL framework that facilitates the construction of deep networks, explicitly reformulating the layers so that they learn from residual functions. Regarding its application, in (Zhu et al., 2019), a residual care network is proposed for the automatic inspection of the weld joint quality; another example is shown in (Helwan et al., 2021), where a DL system is developed by using a residual network type ResNet-50 to classify bananas, discriminating them into healthy or defective.

Other works (Szegedy et al., 2017) show that it is possible to significantly speed up the training of neural networks using residual connections, under architectures known as: Inception-ResNet that have shown good classification results. In (Song et al., 2019), the authors propose a residual squeeze-and-excitation network applied to visual inspection to verify the quality of adhesives on battery cell surfaces, which is achieved with the design of a compact architecture design and an attention mechanism, which provides good results with few training samples.

Transfer Learning

TL is a technique in which a model, previously trained in a given task, is reused in another (usually) related one. The main advantages of TL are that large amounts of training data are not required and that it can help reduce hardware costs for the training stage, which is often computationally expensive (Smith et al., 2021). The objective of TL is to improve the performance taking advantage of the knowledge obtained in a different task (Hafemann et al., 2015), in order to overcome representational deficiencies due to insufficient or uninformative examples in the datasets (Abubakar et al., 2020).

Regarding its application in visual inspection, in (Ren et al., 2018) the authors propose a generic approach that requires small training sets. First, a classifier is built on the feature representation obtained from the patches of each image and, afterwards, those features are transferred to a previously trained DL network. The prediction is then made by convolving the trained classifier on the input image. The system was successfully tested for flaw detection in sheet steel, arc welds, and wood sheets.

Most of the importance of TL consists in using systems in completely different domains from the one used during its training. In (Kim et al., 2017), for example, the authors develop a system for inspecting texture images and successfully transfer the weights obtained from a source network trained with images extracted from the ImageNet dataset. Similarly, in (Hafemann et al., 2015), the authors propose a TL method for different texture classification problems, in order to take advantage of the CNN architecture in problems with smaller datasets, reporting good results on small datasets with macroscopic images used for forest species recognition.

Data Augmentation

A frequent problem that arises in visual inspection is the lack of images to carry out an adequate training process, either due to the difficulty of the acquisition or because there are not enough images of the events to be classified or the cost of labeling them is too high. DA increases the amount and variety of images by manipulating them in various ways, e.g., by rotating, flipping, shifting, resizing, and randomly cropping (Takahashi et al., 2020). In (Perez and Wang, 2017), a comparison of multiple solutions to the problem of the lack of images in classification scenarios is made, proposing a method called neural augmentation; likewise, in (Chen et al., 2018), an application of the ensemble approach with DA is proposed for the recognition of defects in steel surfaces that, unlike the approaches based on a single model, combines several DCNN models and uses DA to grow the defect dataset. Networks such as ResNet-32, WRN-28-10 and WRN-28-20 are included in this approach; they are individually trained and then combined together with an averaging strategy for a joint classification of the defect.

METHODS

A set of solutions was explored, which initially require an image processing stage, followed by the application of different DL approaches that would allow finding an adequate solution to the classification problem of the glass surface defects.

Various types of CNN models were tested, ranging from a CNN baseline model, to deep residual networks in which the TL technique is applied; additionally, three variants of the GI dataset were built, the first one with a random partition of the training and test sets, the second one to guarantee the participation of all the defect categories in both the training and test stages and, finally, one that seeks to balance the amount of images in the different categories. The proposed experiments are defined under the combination of the different variants of network model, dataset, TL and DA techniques; they are illustrated in Figure 1 and summarized in Table 1.

Dataset Preprocessing

The images of the GI dataset were downloaded from (Deltamax Automazione, 2016), which are part of the RISOLVI project (Plotegher et al., 2016). The dataset contains 660 defects in glass sheets with their respective labels for a total of 1160 images; some examples are shown in Figure 2. Image sizes vary between 61 and 346 pixels wide and between 39 and 455 pixels long. Each image has a unique class label according to the defect category it contains. The defects found in the dataset differ from those reported in the companion paper (Plotegher et al., 2016), not only in the number of images per class, but also in the classes themselves; that is, some of the classes reported in (Plotegher et al., 2016) are not present in (Deltamax Automazione, 2016) and vice-versa. Table 2 shows the nominal (according to (Plotegher et al., 2016)) and real (according to (Deltamax Automazione, 2016)) distributions of the images contained in the dataset.

DL-based image recognition systems require the input sample sizes to be equal; therefore, it was necessary to resize all the images and, given the requirements of some models, especially those with TL. A size of 224 x 224 pixels was used with the three channels (RGB) in all the experiments in order to homogenize the test conditions. For all the resulting images, pixels that initially had values between 0 and 255, were standardized to the range (0,1). Concerning the labels, they were defined with numbers, according to the correspondence shown in Table 3.

Figure 1. A tree diagram illustrating the ten considered experiments, according to different combinations of dataset condition (original, guaranteed, balanced), inclusion or exclusion of DA in training and model design (either baseline or transferred)

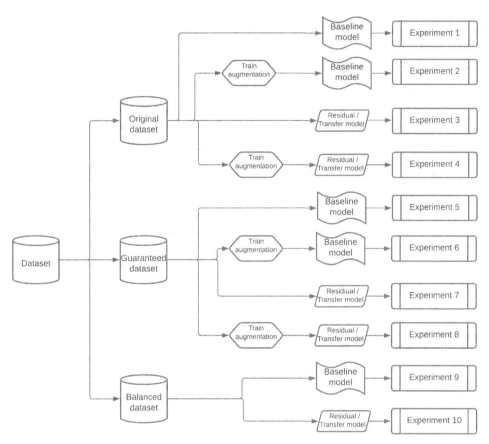

For Experiments 1 to 4, the collection of images was randomly divided into two sets: training and test, in a proportion of 70% and 30% respectively; that partition is referred hereafter as the original dataset, having 812 images for training and 348 for testing.

In contrast, in the other experiments, the participation of all the categories in both training and test phases was ensured in order to deal with the class imbalance problem. Notice that, as seen in Table 3, some categories have a very low number of images (*other* with 2 images and *high_contrast* with 4 images) while other classes are better represented (*stain* has 239 images and *scratch light* has 201 images).

According to the above-mentioned conditions, three variants of the dataset were built, as follows:

Table 1. A summary of the ten considered experiments, according to different combinations of dataset condition (original, guaranteed, balanced), inclusion or exclusion of DA and model design (either baseline or transferred)

#	Dataset	Augmentation	CNN design
1	Original	None	Baseline
2	Original	In training	Baseline
3	Original	None	RL/TL
4	Original	In training	RL/TL
5	Guaranteed	None	Baseline
6	Guaranteed	In training	Baseline
7	Guaranteed	None	RL/TL
8	Guaranteed	In training	RL/TL
9	Balanced	Differential	Baseline
10	Balanced	Differential	RL/TL

Original Dataset

This set is composed by the 1160 original images where the partition into training and test is carried out at random, in a proportion 70% - 30%, just before each experiment. This dataset was used in Experiments 1, 2, 3 and 4.

Guaranteed Dataset

A version of the dataset to avoid that some of the categories do not have representative images in the training set. It is composed by 1160 images which are previously

Figure 2. Examples of defects, with their corresponding labels, from the GI dataset

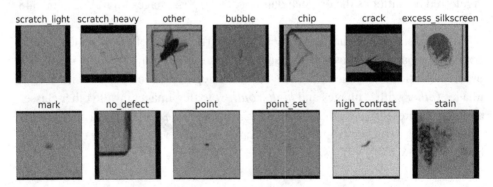

Table 2. Categories and number of defects, nominal (according to (Plotegher et al., 2016)) and real (according to (Deltamax Automazione, 2016)), in the GI dataset

Defect	Nominal	Real
scratch_light	103	103
scratch_heavy	51	51
other	1	1
bubble	52	52
chip	81	82
crack	14	14
excess_silkscreen	0	22
mark	74	74
no_defect	12	13
point	57	57
point_set	107	107
high_contrast	0	3
stain	82	81
printing_error	26	0
Total	**660**	**660**

Table 3. Numerical labels and number of images per category in the GI dataset

Label	Defect	# of images
0	scratch_light	201
1	scratch_heavy	103
2	other	2
3	bubble	70
4	chip	133
5	crack	40
6	excess_silkscreen	36
7	mark	132
8	no_defect	13
9	point	63
10	point_set	124
11	high_contrast	4
12	stain	239
Total		**1160**

distributed in 70% for training and 30% for testing, in such a way that the presence of images of all categories in both the training partition and the test partition is guaranteed, resulting in 812 images for training and 348 images for testing. This dataset was used in Experiments 5, 6, 7 and 8.

Balanced Dataset

This variant of the dataset is aimed to ensure the presence of images from all categories in the training and test sets, as well as to balance the number of images in the categories. The DA process was carried out in differential form using the following parameters: rotation range up to 20 degrees, zoom range of maximum 15%, lateral and superior translation of up to 10% and the possibility of making horizontal and vertical mirror operations (flip). After the augmentation, whose goal was to reach about 400 images per category, the training set size grew from 812 to 4841 images, while the test set was kept with 348 images without augmentation. The composition per class of the balanced dataset is shown in Table 4. This variant of the dataset was used in Experiments 9 and 10.

Baseline Model

As a first option, a CNN was designed, which was based on the Keras repository of examples (Francois Chollet, 2021) and modified according to the requirements of the problem in such a way that it allowed to consider different configurations, architectures, number and types of layers, as well as the adjustment of the different parameters and hyperparameters required for their training.

The main properties that were taken into account to develop and adapt the model include the use of convolutional layers as the core of CNNs and the main tool for image processing, as well as being able to experiment with different types of layers including convolutional, dense, dropout, pooling and flatten versions; in addition to exploring options for the number of filters and the number of layers. After several trials, a CNN architecture with the following specifications was found as the best model built from scratch: up to 19 layers, starting with a convolutional layer that receives the images with a size of 224 x 224 pixels x 3 channels, applying 32 filters of size 3 x 3, in order to extract information and reduce its dimension, followed by layers composed by two sequences of convolutional ones, with normalization, which are repeated for 64, 128, 256, 512, 728 and 1024 filters and are aimed to reduce the size of the image and increase the information until obtaining images with a size of 7 x 7; afterwards, a global grouping layer is applied to reduce the size of the image, followed by a regularization using a dropout layer that seeks to ignore 50% of the neurons. The resulting model, see Table 5, was used in Experiments 1, 2, 5, 6 and

Table 4. Number of images per category for training (original and augmented) and test

Defect	Train orig.	Train augm.	Test
scratch_light	141	423	60
scratch_heavy	72	360	31
other	1	400	1
bubble	49	392	21
chip	93	372	40
crack	28	392	12
excess_silkscreen	25	400	11
mark	92	368	40
no_defect	9	396	4
point	44	396	19
point_set	87	435	37
high_contrast	3	399	1
stain	168	504	71
Total	**812**	**4841**	**348**

9 and includes a dense layer that connects all the neurons of the previous layer with an array of probabilities whose dimension is the number of classes.

Transfer Learning Model with Residual Learning

As a second option, some of the different RL-based network architectures that use TL were selected which, according to the literature (Abubakar et al., 2020), offer the best results for image classification problems. The following ones stand out: ResNet50, ResNet101, ResNet152 and InceptionResNet (Xiao et al., 2018).

After having performed exploratory tests with the above-mentioned architectures and according to the recommendations given in (Hershey et al., 2017, Kornblith et al., 2019), a CNN with the Inception ResNet —pre-trained in the ImageNet dataset— was used but modified by replacing the last fully connected layer with a classifier block. This block includes a global average grouping layer to minimize overfitting, by decreasing the number of parameters, a dropout layer to deactivate a percentage of neurons, and a final dense layer as a softmax classifier, where the value of the normalized probability of the thirteen categories is computed. This network architecture was used for Experiments 3, 4, 7, 8 and 10.

Table 5. CNN architecture of the baseline model as reported by Keras

Layer (type)	Model
Input	(None, 224, 224, 3)
Conv2D	(None, 112, 112, 32)
Batch normalization	(None, 112, 112, 32)
Conv2D 1	(None, 112, 112, 64)
Batch normalization 1	(None, 112, 112, 64)
Separable conv2D	(None, 112, 112, 128)
Batch normalization 2	(None, 112, 112, 128)
Separable conv2D 1	(None, 112, 112, 128)
Batch normalization 3	(None, 112, 112, 128)
Max pooling 2D	(None, 56, 56, 128)
Separable conv2D 2	(None, 56, 56, 128)
Batch normalization 4	(None, 56, 56, 128)
Separable conv2D 3	(None, 56, 56, 256
Batch normalization 5	(None, 56, 56, 256
Max pooling 2D 1	(None, 56, 56, 256
Separable conv2D 4	(None, 28, 28, 512)
Batch normalization 6	(None, 28, 28, 512)
Separable conv2D 5	(None, 28, 28, 512)
Batch normalization 7	(None, 28, 28, 512)
Max pooling 2D 2	(None, 14, 14, 512)
Separable conv2D 6	(None, 14, 14, 728)
Batch normalization 8	(None, 14, 14, 728)
Separable conv2D 7	(None, 14, 14, 728)
Batch normalization 9	(None, 14, 14, 728)
Max pooling 2D 3	(None, 7, 7, 728)
Separable conv2D 8	(None, 7, 7,1024)
Batch normalization 10	(None, 7, 7,1024)
Global average pooling2D	(None, 1024)
Dropout	(None, 1024)
Dense	(None, 13)

Data Augmentation

DA can improve the training phase of a CNN since, providing additional training examples artificially derived from the original ones, the DL models are regularized helping therefore to avoid their overfitting (Mormont et al., 2018, Voulodimos et al., 2018). Taking into account the reduced cardinality of some categories in the dataset and, especially the imbalance between them, it was decided to apply two types of augmentation processes: the first one seeking to increase the number of images at training time; the second one, carried out in a previous process, aimed to balance the number of images between the categories making use of the differential DA technique (Takase et al., 2021), which firstly guarantees that all the categories are represented in the training set, computing later the amount of additional images that are required. The first type of DA was applied to Experiments 2, 4, 6 and 8, while the differential augmentation was applied to Experiments 9 and 10.

RESULTS

Python 3.8.5 was used, with Tensorflow 2.4.1 under Keras version 2.4.0. As test platforms, Google cloud servers under Collaboratory and a DELL PowerEdge T630 server with a Tesla K40c GPU were used.

Experimental Setup

Details of the (hyper)parameter tuning, model training and performance estimation are given below.

(Hyper)parameter Tuning and Model Training

Once the models with either a baseline design or using RL/TL were defined, its operation was checked by selecting and tuning the (hyper)parameters. A review of the different optimization functions that can be used in this type of problem was made, finding that the most appropriate ones are: *Stochastic Gradient Descent*, *Adaptive Moment Estimation* and *Root Mean Square Propagation* (RMSprop) (De et al., 2019). Similarly, regarding the loss function, the following options were taken into account: *Mean Squared Error* and *Categorical Cross-entropy* (Chollet, 2017). Finally, as performance metrics, the use of: *Accuracy* and *Categorical Accuracy* were explored. After the exploratory experiments, the best results were found using *RMSprop*, *Categorical Cross-entropy* and *Accuracy*.

The following considerations regarding the parameters were taken into account to train the models: a maximum of 100 epochs, including an early stopping option when the validation process exceeded 10 epochs without improvement and a batch size of 32 for which the best results were obtained.

Table 6. Accuracies of the experiments. Minimum, maximum and average values are reported along with the standard deviation computed from the ten repetitions of each experiment

# Exp.	Model	Min.	Max.	Avg.	STD
1	Baseline	0.6868	0.7557	**0.7129**	0.0225
2	Baseline	0.6839	0.7902	0.7394	0.0303
3	RL/TL	0.7500	0.8592	0.8112	0.0358
4	RL/TL	0.7557	0.8649	0.8316	0.0296
5	Baseline	0.6983	0.7529	0.7181	0.0184
6	Baseline	0.6954	0.7730	0.7425	0.0287
7	RL/TL	0.8075	0.8391	0.8207	0.0115
8	RL/TL	0.7931	0.8621	0.8336	0.0169
9	Baseline	0.7414	0.7931	0.7698	0.0156
10	RL/TL	0.8218	0.8592	**0.8365**	0.0119

Performance Evaluation

Ten repetitions for each experiment were made to take into account the stochastic nature of the CNN models. Accuracies were computed and reported in terms of the smallest, largest and average values, together with the corresponding standard deviation; additionally, to better illustrate and analyze the results, the cumulative confusion matrix of some of the experiments is presented, as well as a classification report that includes precision, recall and F1-score for each category, and finally the global metrics (macro and average) for precision, recall and F1-score.

To carry out the experiments with the original dataset, a distribution of the training and test sets was made just before the training for each execution: in contrast, for the experiments with the guaranteed and balanced datasets, ten versions of each dataset were previously built, in which the images were randomly distributed, generated, stored in files and read when training the models.

Experimental Results and Discussions

As global results, accuracies obtained in the ten experiments are displayed in Table 6 along with their corresponding error bars in Figure 3.

In a first analysis of the global results, it can be noted that the models have a significant variability in performance, exhibiting differences that, on average, ranged from 0.7129 ± 0.0225 (in Experiment 1), to 0.8365 ± 0.0119 (in Experiment 10).

Figure 3. Accuracy error bars per experiment. Experiments 1, 2, 5, 6, 9 correspond to the baseline model and Experiments 3, 4, 7, 8 and 10 correspond to the transfer model; Experiments 1, 2, 3, 4 use the original dataset, Experiments 5, 6, 7, 8 use the guaranteed dataset and Experiments 9, 10 use the balanced dataset; finally, Experiments 2, 4, 6, 8 use augmentation in training

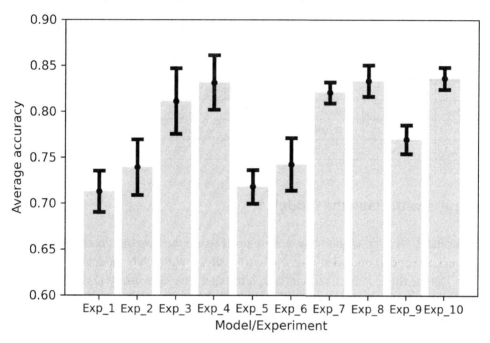

Analyzing in more detail, overall performance metrics including macro and average precision, recall and F1-score metrics are presented in Table 7.

From Figure 3, it can be seen that between pairs of experiments whose only difference is whether DA is applied or not (Experiments 2-1, 4-3, 6-5, 8-7), improvements in the average performance are observed but they are not significant under the simple criterion of non-overlapping $Avg \pm SD$ intervals. Regarding the F1 macro score, notice that the largest values are observed in Experiments 7, 9 and 10. On the contrary, considering the weighted metrics, which could be more useful given the imbalance between the classes, the results obtained in Experiments 3, 4, 7, 8, and 10 using transfer model stand out, having values above 0.8 in the weighted F1-score.

Since two different models (baseline and RL/TL) were used, further detailed analyses of the experimental results are separately discussed below for those two cases.

Table 7. Overall performance metrics per experiment

# Exp.	Macro pre.	Macro rec.	Macro F1	Weigh pre.	Weigh rec.	Weigh F1
1	0.6144	0.5604	0.5759	0.7138	0.7129	0.7076
2	0.6084	0.5732	0.5792	0.7356	0.7394	0.7324
3	0.7137	0.6689	0.6823	0.8114	0.8112	**0.8072**
4	0.6955	0.6693	0.6767	0.8243	0.8316	**0.8249**
5	0.6354	0.5705	0.5914	0.7198	0.7181	0.7126
6	0.6804	0.5921	0.6063	0.7475	0.7425	0.7330
7	0.7200	0.6907	**0.7004**	0.8179	0.8207	**0.8168**
8	0.7030	0.6741	0.6853	0.8292	0.8336	**0.8295**
9	0.7118	0.7099	**0.7083**	0.7699	0.7698	0.7687
10	0.8199	0.7331	**0.7526**	0.8372	0.8365	**0.8354**

Results with Baseline Model

Experiments 1, 2, 5, 6 and 9 were developed using the baseline model, allowing to check the operation and ability to classify defects with CNN models having less layers on the three dataset variants and, alternately, making use of DA techniques.

Figure 4 shows the evolution of the training and test accuracies with respect to the epochs in Experiments 1 and 2, particularly for the runs where the maximum test accuracy was obtained (0.7557 and 0.7902 respectively), finding that the training accuracy is almost 0.1 higher than the test accuracy for Experiment 1, showing that the lack of images harms generalization; in contrast, curves in Experiment 2 are similar, showing that the network learns better and is able to generalize.

When reviewing the cumulative classification report of Experiment 2 (see Table 8), it can be seen that *stain* is the category that gets the best results, with an F1-score of 0.8709, while the worst results are obtained in categories such as *other* and *high_contrast*, where the number of images is less than 10. This implies that for some of the ten runs there were no available images during training, affecting the classifier results since this category was not trained and, therefore, some performance measures get 0.0 for those classes.

For Experiment 5, which uses the guaranteed dataset on the baseline model, an average accuracy of 0.7181 ± 0.0184 was reached, with a small improvement of 0.0052 with respect to Experiment 1. This improvement is explained by the fact that, in eleven of the thirteen categories the number of images to train was the same as in the previous experiment, but in *other* and *high_contrast*, the participation of images is ensured, guaranteeing at least one image per class. The next step in Experiment 6,

Figure 4. Learning curve - accuracy vs epochs

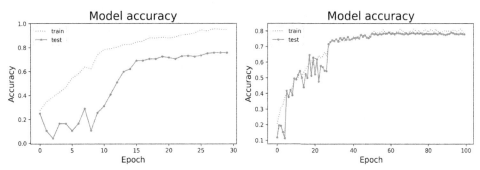

Experiment 1 : baseline model without DA   Experiment 2 : baseline model with DA

was training the model with a guaranteed dataset but applying DA, thus achieving an average accuracy of 0.7425 ± 0.0287.

Figure 5 shows the evolution of the training and test accuracies of Experiments 5 and 6, presented for the runs in which they obtained the maximum accuracy (0.7529 and 0.7730 respectively). Notice that the accuracy of the training set is about 0.2 higher than the accuracy of the test set for Experiment 5, showing that the lack of images harms generalization and, that for Experiment 6 the curves come together, showing that the network is capable of learning and generalizing reasonably well.

Table 8. Cumulative classification report for Experiment 2

Category	Precision	Recall	F1-score	Support
scratch_light	0.6788	0.7598	0.7170	587
scratch_heavy	0.5350	0.5733	0.5535	307
other	**0.0000**	**0.0000**	**0.0000**	7
bubble	0.8861	0.8524	0.8689	210
chip	0.8175	0.8281	0.8228	384
crack	0.7053	0.5630	0.6262	119
excess_silkscreen	0.5417	0.1238	0.2016	105
mark	0.6738	0.6429	0.6580	392
no_defect	0.7667	0.6571	0.7077	35
point	0.7336	0.8441	0.7850	186
point_set	0.6995	0.7367	0.7176	376
high_contrast	**0.0000**	**0.0000**	**0.0000**	8
stain	0.8715	0.8704	**0.8709**	764

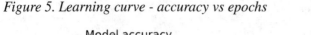

Figure 5. Learning curve - accuracy vs epochs

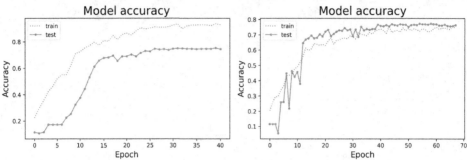

Experiment 5 : baseline model without DA   Experiment 6 : baseline model with DA

It is important to analyze the cumulative confusion matrix from Experiment 6 that is presented in Table 9, as well as the classification report in Table 10, where the best results achieved so far are shown, finding that some categories of defects such as *scratch_light* and *scratch_heavy* are prone to be confused with each other, which explains their misclassifications: 42 images (7.0%) of *scratch_light* that are classified as *scratch_heavy*, and 71 images (11.8%) of *scratch_heavy* that are classified as *scratch_light*. Notice also that 49 images of *scratch_light* (8.2%) are classified as *point_set* which can be visually similar.

Additionally, for categories such as *other* and *high_contrast* it can be seen that with a total of 10 images analyzed in ten experiments, they only had one image for training and 1, 2 or 3 images respectively to test; moreover, DA does not guarantee enough images in those categories, which is reflected in the low results of the combined metric F1-score with 0.0 and 0.3750 respectively.

The categories that show the best results in terms of precision are *excess_silkscreen* with 0.9167 and *bubble* with 0.914 while, in terms of recall, they are *chip* with 0.87 and *stain* with 0.8592. Regarding F1-score, the best results correspond to *bubble* with 0.8586 and *stain* with 0.8543, which shows that the classifier behaves very well in those classes. The largest difference between precision and recall is observed for *excess_silkscreen* (0.9167 vs. 0.1), indicating that the model does not detect this class very well, but when it does, its prediction is quite reliable.

The categories with the worst precision and recall in Experiment 6 are *other* (0.0 and 0.0 respectively) and *high_contrast* (0.5 and 0.3 respectively). Remember that these classes are the ones with the less number of images for training, implying that the model is unable to learn them correctly.

In Experiment 9, that uses the balanced dataset, an accuracy of 0.7698 ± 0.0156 was obtained (refer again to Table 6), which is a slight but not significant increase of 0.0273 with respect to the best results of Experiment 6.

Table 9. Cumulative confusion matrix for Experiment 6

Category	0	1	2	3	4	5	6	7	8	9	10	11	12	Total
0 - scratch_light	468	42	0	0	6	0	0	12	0	1	49	0	22	600
1 - scratch_heavy	71	184	0	1	14	5	0	13	0	4	12	0	6	310
2 - other	0	0	0	0	10	0	0	0	0	0	0	0	0	10
3 - bubble	3	0	0	170	0	0	0	12	0	24	1	0	0	210
4 - chip	4	2	0	0	348	9	1	2	7	0	2	0	25	400
5 - crack	2	27	0	0	11	73	0	5	0	0	2	0	0	120
6 - excess_silkscreen	1	3	0	1	16	8	11	32	0	5	4	0	29	110
7 - mark	23	35	0	7	8	0	0	275	0	19	14	1	18	400
8 - no_defect	5	0	0	0	11	0	0	0	23	0	0	0	1	40
9 - point	4	0	0	5	0	0	0	20	0	149	10	2	0	190
10 - point_set	54	8	0	2	3	0	0	9	0	17	270	0	7	370
11 - high_contrast	1	0	0	0	0	0	0	1	0	5	0	3	0	10
12 - stain	33	10	0	0	16	0	0	22	0	0	19	0	610	710

The following observations can be made from a further analysis of the report in Table 10: the best results are obtained for *bubble*, *chip* and *stain*, whose F1-scores

Table 10. Cumulative classification report of Experiments 6 and 9

Category	Experiment 6			Experiment 9			
	Precision	Recall	F1-score	Precision	Recall	F1-score	Support
scratch_light	0.6996	0.7800	0.7376	0.7320	0.7467	0.7393	600
scratch_heavy	0.5916	0.5935	0.5926	0.6536	0.6452	0.6494	310
other	**0.0000**	**0.0000**	0.0000	0.6667	0.6000	0.6316	10
bubble	**0.9140**	0.8095	**0.8586**	0.9146	0.8667	**0.8900**	210
chip	0.7856	**0.8700**	0.8256	0.8722	0.8700	**0.8711**	400
crack	0.7684	0.6083	0.6791	0.7500	0.7750	0.7623	120
excess_silkscreen	**0.9167**	**0.1000**	0.1803	0.4425	0.4545	**0.4484**	110
mark	0.6824	0.6875	0.6849	0.7311	0.6050	0.6621	400
no_defect	0.7667	0.5750	0.6571	0.7234	0.8500	0.7816	40
point	0.6652	0.7842	0.7198	0.7285	0.8474	0.7835	190
point_set	0.7050	0.7297	0.7171	0.7538	0.8027	0.7775	370
high_contrast	**0.5000**	**0.3000**	0.3750	0.4286	0.3000	**0.3529**	10
stain	0.8496	**0.8592**	**0.8543**	0.8565	0.8662	**0.8613**	710

are higher than 0.8, suggesting therefore that they can be classified with reliability. In contrast, the lowest F1-scores (0.3529 and 0.4484) correspond to *high_contrast* and *excess_silkscreen*, indicating that those categories are unreliable and difficult to classify.

Training and test learning curves and cumulative confusion matrix for Experiment 9 are shown in Figure 6 and Table 11 respectively. It can be seen that the model is not good enough to generalize since the difference between training and test accuracies is greater than 0.1; therefore, a balanced dataset does not compensate for the limited variety of images in some of the categories, which is corroborated with the F1-score for categories like *high_contrast* which only reaches 0.3529. Another important factor is the morphological affinity between some of the categories seen in Figure 2, finding that *scratch_light, scratch_heavy* and *point_set* tend to be confused with each other, as well as with *mark* that morphologically shares features with other categories; in this case, the classifier ends up discriminating 400 images in 11 of the 13 available categories.

Results with Transfer Learning for a Residual Learning Model

Experiments 3, 4, 7, 8 and 10 take advantage of TL on a residual CNN model. Accuracies are presented in Table 6, where Experiment 10 obtained the best average results with 0.8365 ± 0.0119.

Regarding Experiment 3, an average accuracy of 0.8112 ± 0.0358 for the original dataset was obtained, which represents an improvement of 0.0414 with respect to the best result with the baseline model. Subsequently, in Experiment 4, the capacity of the network with DA was tested, aiming to better characterize the categories, with a result of 0.8316 ± 0.0296 in accuracy. Note that, according to Figure 3, results of using DA in Experiments 3 and 4 do not overlap with the results of Experiments 1 and 2; therefore, using DA can be said to be significantly useful.

Classification results for Experiment 4 are illustrated in its cumulative confusion matrix in Table 12, and the corresponding classification report in Table 13. According to the cumulative confusion matrix, *scratch_light, scratch_heavy,* as well as *point* and *point_set* continue to be confused. Categories with a high recall and F1-score values, greater than 0.8, such as *scratch_light, bubble, chip, crack, no_defect, point, point_set* and *stain*, show that they can be classified with good reliability, while in *other* and *high_contrast* which have a low number of images, it can be seen that, it is not guaranteed the availability of more than one image to train, which is noticed in the null recall for the two categories.

For Experiment 7, the guaranteed dataset was tested in order to compare its behavior against the TL model, obtaining an average accuracy of 0.8207 ± 0.0115 which is 0.0095 better than the one obtained with the original dataset. The explanation

Figure 6. Learning curve - accuracy vs epochs of Experiment 9

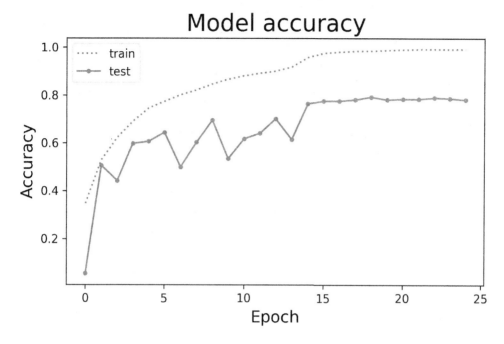

Table 11. Cumulative confusion matrix for Experiment 9

Category	0	1	2	3	4	5	6	7	8	9	10	11	12	Total
0 - scratch_light	448	53	0	0	6	1	3	18	1	4	43	0	23	600
1 - scratch_heavy	61	200	0	0	9	10	4	8	0	3	8	0	7	310
2 - other	0	0	6	0	1	0	2	0	0	0	0	0	1	10
3 - bubble	3	0	0	182	1	0	7	7	0	7	3	0	0	210
4 - chip	9	7	2	0	348	7	6	2	6	0	1	0	12	400
5 - crack	0	10	0	0	9	93	1	6	0	0	0	0	1	120
6 - excess_silkscreen	0	1	1	1	8	3	50	17	0	5	1	0	23	110
7 - mark	19	21	0	9	6	8	28	242	0	27	14	3	23	400
8 - no_defect	3	0	0	0	0	0	0	0	34	0	0	0	3	40
9 - point	1	1	0	6	0	0	0	11	0	161	9	1	0	190
10 - point_set	37	7	0	1	1	2	2	5	0	8	297	0	10	370
11 - high_contrast	0	0	0	0	0	0	0	1	0	6	0	3	0	10
12 - stain	31	6	0	0	10	0	10	14	6	0	18	0	615	710

Table 12. Cumulative confusion matrix for Experiment 4

Category	0	1	2	3	4	5	6	7	8	9	10	11	12	Total
0 - scratch_light	520	32	0	0	4	0	0	14	0	0	32	0	13	615
1 - scratch_heavy	58	224	0	0	0	2	0	12	0	3	10	0	5	314
2 - other	0	0	0	0	7	0	0	0	0	0	0	0	0	7
3 - bubble	1	0	0	171	2	0	0	12	0	8	2	0	1	197
4 - chip	2	0	0	0	362	5	4	0	2	0	2	0	3	380
5 - crack	7	8	0	0	4	103	1	2	0	0	0	0	0	125
6 - excess_silkscreen	0	6	0	0	15	7	31	27	0	1	3	0	29	119
7 - mark	16	12	0	7	1	4	11	316	0	11	15	0	15	408
8 - no_defect	0	0	0	0	5	0	0	0	38	0	0	0	4	47
9 - point	1	0	0	2	0	0	0	18	0	149	7	1	0	178
10 - point_set	11	5	0	0	0	0	0	3	0	7	341	0	6	373
11 - high_contrast	3	0	0	0	0	0	0	0	0	6	0	0	0	9
12 - stain	22	3	0	0	2	2	7	8	0	0	25	0	639	708

Table 13. Cumulative classification report for Experiment 4

Category	Precision	Recall	F1-score	Support
scratch_light	0.8112	**0.8455**	0.8280	615
scratch_heavy	0.7724	0.7134	0.7417	314
other	0.0000	**0.0000**	0.0000	7
bubble	0.9500	**0.8680**	0.9072	197
chip	0.9005	**0.9526**	0.9258	380
crack	0.8374	**0.8240**	0.8306	125
excess_silkscreen	0.5741	0.2605	0.3584	119
mark	0.7670	0.7745	0.7707	408
no_defect	0.9500	**0.8085**	0.8736	47
point	0.8054	**0.8371**	0.8209	178
point_set	0.7803	**0.9142**	0.8420	373
high_contrast	0.0000	**0.0000**	0.0000	9
stain	0.8937	**0.9025**	0.8981	708

Table 14. Cumulative confusion matrix for Experiment 8

Category	0	1	2	3	4	5	6	7	8	9	10	11	12	Total
0 - scratch_light	514	27	0	0	4	1	0	14	0	0	26	0	14	600
1 - scratch_heavy	59	220	0	0	0	7	1	12	0	0	5	0	6	310
2 - other	0	0	0	0	9	0	1	0	0	0	0	0	0	10
3 - bubble	2	0	0	180	0	0	1	13	0	9	4	0	1	210
4 - chip	8	0	0	0	374	5	2	0	3	0	3	0	5	400
5 - crack	6	7	0	0	5	98	0	4	0	0	0	0	0	120
6 - excess_silkscreen	0	1	0	0	6	4	48	22	0	5	3	0	21	110
7 - mark	16	4	0	5	3	0	12	313	0	13	14	0	20	400
8 - no_defect	0	0	0	0	3	0	0	0	31	0	0	0	6	40
9 - point	0	0	0	5	0	0	0	18	0	155	10	2	0	190
10 - point_set	21	2	0	0	1	2	0	4	0	7	321	0	12	370
11 - high_contrast	0	0	0	0	0	0	0	1	0	9	0	0	0	10
12 - stain	24	3	0	0	1	0	6	12	0	0	17	0	647	710

is that there should only be improvement in those categories that could run out of images such as *other* and *high_contrast*, which have few images; however, in spite of guaranteeing that those categories will remain represented in the training set, they do not contribute much to the overall results.

In Experiment 8, DA was applied on the guaranteed dataset, obtaining an average accuracy of 0.8336 ± 0.0169 (Table 6), which is 0.0013 better than when the data is not augmented (Experiment 7). However, according to Figure 3, the use of DA is not enough since some repetitions of Experiment 8 do not reach the mean of Experiment 7, confirming that by providing additional images during training, results are just slightly improved.

The cumulative confusion matrix and the corresponding classification report for Experiment 8 are presented in Tables 14 and 15, allowing to observe that: despite that *other* and *high_contrast* have images for training in all the repetitions of the experiment, their recalls are 0.0 since it is not possible to guarantee that the augmentation provides enough images to significantly improve the overall performance of the classifier. Confusions persist between *scratch_light*, *scratch_heavy* and *point_set*. Categories such as *chip* and *stain* achieve recalls greater than 0.9, increasing the reliability of the classifier for these categories. All categories except *other*, *excess_silkscreen* and *high_contrast* show high and similar results in precision and recall, which indicates that the model will perform well to classify images in these categories.

Table 15. Cumulative classification report for Experiments 8 and 10

Category	Experiment 8			Experiment 10			Support
	Precision	Recall	F1-score	Precision	Recall	F1-score	
scratch_light	0.7908	0.8567	0.8224	0.8376	0.8250	0.8312	600
scratch_heavy	0.8333	0.7097	0.7666	0.8147	0.7516	0.7819	310
other	0.0000	**0.0000**	0.0000	**1.0000**	**0.2000**	**0.3333**	10
bubble	0.9474	0.8571	0.9000	0.9118	0.8857	0.8986	210
chip	0.9212	**0.9350**	0.9280	0.9474	0.9450	0.9462	400
crack	0.8376	0.8167	0.8270	0.7744	0.8583	0.8142	120
excess_silkscreen	0.6761	0.4364	0.5304	**0.5377**	**0.5182**	**0.5278**	110
mark	0.7579	0.7825	0.7700	0.7807	0.7475	0.7637	400
no_defect	0.9118	0.7750	0.8378	0.9429	0.8250	0.8800	40
point	0.7828	0.8158	0.7990	0.7824	0.7947	0.7885	190
point_set	0.7965	0.8676	0.8305	0.7650	0.8622	0.8107	370
high_contrast	0.0000	**0.0000**	0.0000	**0.6667**	**0.4000**	**0.5000**	10
stain	0.8839	**0.9113**	0.8974	0.8979	0.9169	0.9073	710

Table 16. Cumulative confusion matrix for Experiment 10

Category	0	1	2	3	4	5	6	7	8	9	10	11	12	Total
0 - scratch_light	**495**	35	0	0	3	0	2	11	0	0	43	0	11	600
1 - scratch_heavy	43	233	0	0	3	13	2	7	0	0	6	0	3	310
2 - other	0	0	2	0	5	0	3	0	0	0	0	0	0	10
3 - bubble	0	0	0	186	2	0	3	14	0	3	2	0	0	210
4 - chip	9	0	0	0	378	3	3	1	2	0	0	0	4	400
5 - crack	4	2	0	0	4	103	2	5	0	0	0	0	0	120
6 - excess_silkscreen	0	0	0	0	4	4	57	18	0	6	4	0	17	110
7 - mark	10	8	0	10	0	2	18	299	0	21	14	1	17	400
8 - no_defect	1	0	0	0	0	0	0	0	33	0	0	0	6	40
9 - point	1	2	0	7	0	0	1	18	0	151	9	1	0	190
10 - point_set	13	4	0	0	0	5	4	2	0	7	319	0	16	370
11 - high_contrast	0	0	0	1	0	0	0	0	0	5	0	4	0	10
12 - stain	15	2	0	0	0	3	11	8	0	0	20	0	651	710

Figure 7. Learning curve - accuracy vs epochs for Experiment 10

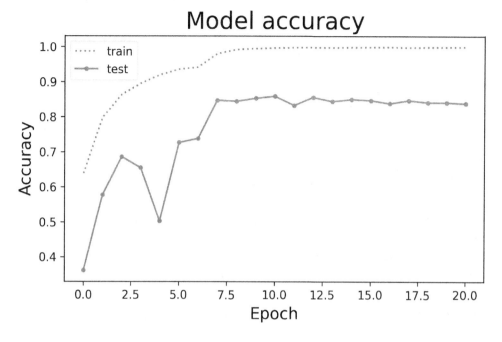

Finally, Experiment 10 that uses the balanced dataset, gets an average accuracy of 0.8365 ± 0.0119 (Table 6), the highest achieved, which is 0.0029 higher than the one of Experiment 8, with the guaranteed dataset. The following observations can be made from the analysis of the confusion matrix (Table 16) and the classification report (Table 15) of Experiment 10:

All categories except *other*, *excess_silkscreen* and *high_contrast* show results higher than 0.75 in precision, recall and F1-score. Categories such as *other* and *high_contrast*, that had null recalls in previous experiments, now reach values of 0.3333 and 0.5 respectively, giving the possibility of correctly classifying some of their images. The appreciations regarding the affinity of some categories such as *scratch_light*, *scratch_heavy* and *point_set* are maintained.

Figure 7, shows training and test curves for Experiment 10. It can be seen that, despite reaching 1.0 accuracy in training, the model is not sufficiently good to generalize everything, since the maximum value reached in test is 0.8365 ± 0.0119. In spite that a balanced version of the dataset is used in this experiment, some of the categories remain under-represented. Another important factor that limits the test accuracy is the morphological affinity that exists between some of the categories, finding that *scratch_light*, *scratch_heavy* and *point_set* share many of their physical features. It is also important to highlight the performance similarity of the experiments that include DA; see again the results for Experiments 3, 4, 7, 8 and 10 in Table 6 and Figure 3. Notice that the accuracy means range from 0.8112 in Experiment 3, reaching up to 0.8365 in Experiment 10.

Reliability of the Original Class Labels

A manual examination of the images and their original class labels was carried out, finding that many of them contained more than one defect or could be mislabeled. Figure 8 shows exemplar groups of images that are visually very similar and, therefore, prone to be easily confused or even ambiguously or incorrectly labeled by the experts.

Another check of the dataset complexity is presented in Figure 9, where the confidences of the predicted labels for some images are shown using the best run from Experiment 10. Notice, that an image of class *scratch_heavy* was mainly predicted as *crack* (category 5) with a posterior of about 1.0; similarly, a *mark* image is predicted mostly as *bubble* (category 3) with a posterior of 1.0; and the following figures show similar cases between the classes *bubble* and *mark*, *excess_silkscreen* and *stain*, *mark* and *stain*, and between *scratch_light* and *point_set*, showing the affinity between them or potential inconsistencies in the original labeling.

FUTURE RESEARCH DIRECTIONS

As recently pointed out in (Smith et al., 2021), an emerging trend in machine vision is the combination of conventional image recognition techniques with DL models, such that the latter are improved either in terms of classification performance, computational complexity, energy consumption or speed of operation. Therefore, future work may include the hybridization of DL models through the incorporation of conventional but recent techniques such as those of learning with inexact supervision (Zhou, 2017), particularly aimed to improve or simplify the preprocessing stages in complex DL-based systems. In addition, efforts to migrate simulated results to realistic industrial environments are also required; for instance, by using compact and dedicated hardware solutions such as single-board microcontrollers/microcomputers (e.g. Raspberry Pi, Arduino, etc.) and modern programmable logic controllers.

CONCLUSION

A comparative analysis of several CNN-based strategies for classifying defects in glass sheets was presented in this chapter. The comparison included ten experiments that considered different options for dataset preprocessing and CNN architectures,

Figure 8. Examples of similar groups images with ambiguous original labeling

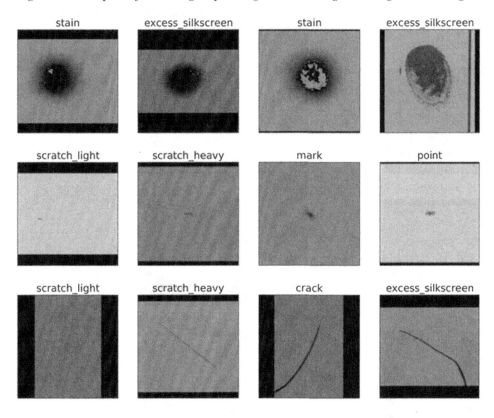

as well as deep learning techniques such as TL, RL, DA and (hyper)parameter settings; in addition, a variety of metrics to measure and interpret the resulting performances were used. The best results were achieved when using RL-based networks, particularly the Inception ResNet architecture along with the combined application of TL and DA techniques, reaching an average accuracy of 0.8365 ± 0.0119, a precision of 0.8372, a recall of 0.8365 and a combined F1-score of 0.8354. Such performances are considered acceptable given the challenging nature of the GI dataset and the potential ambiguity found in the original labeling of its images. The exhaustive performance analysis to understand the behavior of the solutions also included confusion matrices, detailed classification reports, training/test learning curves and confidences of the predicted class labels. Finally, it was clear that the ambiguity found in the original labels of the GI dataset and the under-representation of some of the defect categories are drawbacks that can only be partially overcome by using the most sophisticated CNN models and deep learning techniques.

Figure 9. Posteriors/confidences of the predicted class labels in the last run of Experiment 10. Correspondences between category names and indexes of the predicted labels are the same that were indicated in Table 3

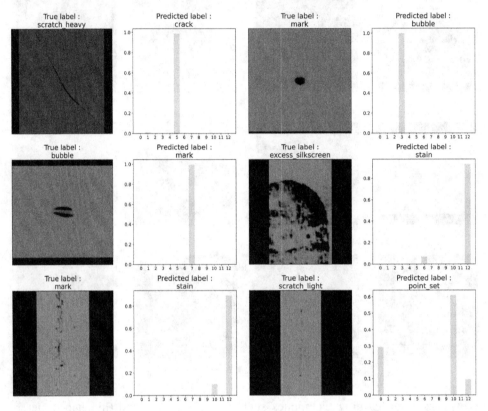

ACKNOWLEDGMENT

This research was supported by Universidad Nacional de Colombia - Sede Manizales. The authors acknowledge Dr. Chiara Corridori, Technical Manager from Deltamax Automazione.

REFERENCES

Abubakar, A., Ajuji, M., & Usman Yahya, I. (2020). Comparison of deep transfer learning techniques in human skin burns discrimination. *Applied System Innovation*, *3*(2).

Cha, Y.-J., Choi, W., Suh, G., Mahmoudkhani, S., & Buyukozturk, O. (2017). Autonomous structural visual inspection using region-based deep learning for detecting multiple damage types. *Computer-Aided Civil and Infrastructure Engineering, 33*(9), 731–747. doi:10.1111/mice.12334

Chen, S. H., & Perng, D. B. (2011). Directional textures auto-inspection using principal component analysis. *International Journal of Advanced Manufacturing Technology, 55*(9-12), 1099–1110. doi:10.100700170-010-3141-1

Chen, W., Gao, Y., Gao, L., & Li, X. (2018). A new ensemble approach based on deep convolutional neural networks for steel surface defect classification. *Procedia CIRP, 72*, 1069–1072. doi:10.1016/j.procir.2018.03.264

Chen, W.-C., & Hsu, S.-W. (2007). A neural-network approach for an automatic LED inspection system. *Expert Systems with Applications, 33*(2), 531–537. doi:10.1016/j. eswa.2006.06.011

Chollet, F. (2017). *Deep Learning with Python* (1st ed.). Manning Publications Co.

Chollet, F. (2021). *Image classification from scratch.* https://keras.io/examples/vision/image_classification_from_scratch/

De, S., Mukherjee, A., & Ullah, E. (2018). *Convergence guarantees for RMSProp and ADAM in non-convex optimization and an empirical comparison to Nesterov acceleration.* https://arxiv.org/abs/1807.06766

Automazione, D. (2016). *GI dataset.* https://www.deltamaxautomazione.it/risolvi/

Deng, J., Dong, W., Socher, R., Li, L., Li, K., & Li, F.-F. (2009). ImageNet: A large-scale hierarchical image database. *2009 IEEE Conference on Computer Vision and Pattern Recognition*, 248–255. 10.1109/CVPR.2009.5206848

Hafemann, L. G., Oliveira, L. S., Cavalin, P. R., & Sabourin, R. (2015). Transfer learning between texture classification tasks using convolutional neural networks. *2015 International Joint Conference on Neural Networks (IJCNN)*, 1–7. 10.1109/IJCNN.2015.7280558

Hatcher, W. G., & Yu, W. (2018). A Survey of Deep Learning: Platforms, Applications and Emerging Research Trends. *IEEE Access: Practical Innovations, Open Solutions, 6*, 24411–24432. doi:10.1109/ACCESS.2018.2830661

He, K., Zhang, X., Ren, S., & Sun, J. (2016). Deep residual learning for image recognition. *2016 IEEE Conference on Computer Vision and Pattern Recognition (CVPR)*, 770–778. 10.1109/CVPR.2016.90

Helwan, A., Ma'aitah, M., Abiyev, R., Uzelaltınbulat, S., & Sonyel, B. (2021). Deep learning based on residual networks for automatic sorting of bananas. *Journal of Food Quality*, *2021*, 1–11. doi:10.1155/2021/5516368

Hershey, S., Chaudhuri, S., Ellis, D. P. W., Gemmeke, J. F., Jansen, A., Moore, R. C., Plakal, M., Platt, D., Saurous, R. A., Seybold, B., Slaney, M., Weiss, R. J., & Wilson, K. (2017). CNN architectures for large-scale audio classification. *2017 IEEE International Conference on Acoustics, Speech and Signal Processing (ICASSP)*. 10.1109/ICASSP.2017.7952132

Kim, S., Kim, W., Noh, Y., & Park, F. C. (2017). Transfer learning for automated optical inspection. *2017 International Joint Conference on Neural Networks (IJCNN)*, 2517–2524. 10.1109/IJCNN.2017.7966162

Kornblith, S., Shlens, J., & Le, Q. V. (2019). Do better ImageNet models transfer better? *2019 IEEE/CVF Conference on Computer Vision and Pattern Recognition (CVPR)*, 2656–2666. 10.1109/CVPR.2019.00277

Lazebnik, S. (2015). Deep convolutional neural networks for hyperspectral image classification. *Journal of Sensors*, *2015*, 258619.

Lin, H. D., & Ho, D. C. (2007). Detection of pinhole defects on chips and wafers using DCT enhancement in computer vision systems. *International Journal of Advanced Manufacturing Technology*, *34*(5-6), 567–583. doi:10.100700170-006-0614-3

Medina, R., Gayubo, F., González, L. M., Olmedo, D., Gómez-García-Bermejo, J., Zalama, E., & Perán, J. (2011). Automated visual classification of frequent defects in flat steel coils. *International Journal of Advanced Manufacturing Technology*, *57*(9-12), 1087–1097. doi:10.100700170-011-3352-0

Mera, C., Orozco-Alzate, M., Branch, J., & Mery, D. (2016). Automatic visual inspection: An approach with multi-instance learning. *Computers in Industry*, *83*, 46–54. doi:10.1016/j.compind.2016.09.002

Mormont, R., Geurts, P., & Marée, R. (2018). Comparison of deep transfer learning strategies for digital pathology. *2018 IEEE/CVF Conference on Computer Vision and Pattern Recognition Workshops (CVPRW)*, 2343–234309. 10.1109/CVPRW.2018.00303

Perez, L., & Wang, J. (2017). *The effectiveness of data augmentation in image classification using deep learning*. ArXiv, abs/1712.04621.

Plotegher, L., Corridori, C., Dolci, F., Andreatta, C., Benini, S., & Devilli, M. (2016). *The GI dataset for glass inspection 1st release (August 2016). Technical report*. Deltamax Automazione Srl.

Ren, R., Hung, T., & Tan, K. (2018). A generic deeplearning-based approach for automated surface inspection. *IEEE Transactions on Cybernetics*, *48*(3), 929–940. doi:10.1109/TCYB.2017.2668395 PMID:28252414

Shorten, C., & Khoshgoftaar, T. (2019). A survey on image data augmentation for deep learning. *Journal of Big Data*, *6*(1), 1–48. doi:10.118640537-019-0197-0

Smith, M. L., Smith, L. N., & Hansen, M. F. (2021). The quiet revolution in machine vision - a state-of-the-art survey paper, including historical review, perspectives, and future directions. *Computers in Industry*, *130*, 103472. doi:10.1016/j.compind.2021.103472

Song, Z., Yuan, Z., & Liu, T. (2019). Residual squeezeand-excitation network for battery cell surface inspection. *2019 16th International Conference on Machine Vision Applications (MVA)*, 1–5.

Szegedy, C., Ioffe, S., Vanhoucke, V., & Alemi, A. A. (2017). Inception-V4, inception-ResNet and the impact of residual connections on learning. In *Proceedings of the Thirty-First AAAI Conference on Artificial Intelligence* (pp. 4278–4284). AAAI Press. 10.1609/aaai.v31i1.11231

Tabernik, D., Šela, S., Skvarč, J., & Skočaj, D. (2019). Deep-learning-based computer vision system for surface-defect detection. In D. Tzovaras, D. Giakoumis, M. Vincze, & A. Argyros (Eds.), *Computer Vision Systems* (pp. 490–500). Springer International Publishing. doi:10.1007/978-3-030-34995-0_44

Takahashi, R., Matsubara, T., & Uehara, K. (2020). Data augmentation using random image cropping and patching for deep CNNs. *IEEE Transactions on Circuits and Systems for Video Technology*, *30*(9), 2917–2931. doi:10.1109/TCSVT.2019.2935128

Takase, T., Karakida, R., & Asoh, H. (2021). Self-paced data augmentation for training neural networks. *Neurocomputing*, *442*, 296–306. doi:10.1016/j.neucom.2021.02.080

Voulodimos, A., Doulamis, N., Doulamis, A., & Protopapadakis, E. (2018). Deep Learning for Computer Vision: A Brief Review. *Computational Intelligence and Neuroscience*, *2018*, 13. doi:10.1155/2018/7068349 PMID:29487619

Weimer, D., Benggolo, A. Y., & Freitag, M. (2015). Context-aware deep convolutional neural networks for industrial inspection. In *Australasian Joint Conference on Artificial Intelligence*. Queensland University of Technology.

Xiao, T., Liu, L., Li, K., Qin, W., Yu, N., & Li, Z. (2018). Comparison of transferred deep neural networks in ultrasonic breast masses discrimination. *BioMed Research International*, *2018*, 1–9. doi:10.1155/2018/4605191 PMID:30035122

Yu, Z., Wu, X., & Gu, X. (2017). Fully convolutional networks for surface defect inspection in industrial environment. In M. Liu, H. Chen, & M. Vincze (Eds.), *Computer Vision Systems* (pp. 417–426). Springer International Publishing. doi:10.1007/978-3-319-68345-4_37

Zhou, T., Ruan, S., & Canu, S. (2019). A review: Deep learning for medical image segmentation using multi-modality fusion. *Array*, *3-4*, 100004. doi:10.1016/j.array.2019.100004

Zhou, Z.-H. (2017). A brief introduction to weakly supervised learning. *National Science Review*, *5*(1), 44–53. doi:10.1093/nsr/nwx106

Zhu, J., Yuan, Z., & Liu, T. (2019). Welding joints inspection via residual attention network. *2019 16th International Conference on Machine Vision Applications (MVA)*, 1–5. 10.23919/MVA.2019.8758040

ADDITIONAL READING

Huang, Y., Qiu, C., Wang, X., Wang, S., & Yuan, K. (2020). A Compact Convolutional Neural Network for Surface Defect Inspection. *Sensors (Basel)*, *20*(7), 1974. Advance online publication. doi:10.339020071974 PMID:32244764

Pedrycz, W., & Chen, S.-M. (2020). *Development and Analysis of Deep Learning Architectures*. Springer. doi:10.1007/978-3-030-31764-5

Roy, S., Balas, V., Samui, P., & D., S. (2019). *Handbook of Deep Learning Applications*. Springer. doi:10.1007/978-3-030-11479-4

Sokolova, M., & Lapalme, G. (2009). A systematic analysis of performance measures for classification tasks. *Information Processing & Management*, *45*(4), 427–437. doi:10.1016/j.ipm.2009.03.002

KEY TERMS AND DEFINITIONS

Accuracy: The number of classifications that a model correctly predicts over the total number of predictions. This measure is used to determine the fraction of class predictions that a model got right.

Class Imbalance: Scenario in which the number of observations belonging to one class is significantly lower than the number of observations belonging to the other classes.

Confusion Matrix: Matrix used to determine the performance of the classification models. It shows the predicted and actual class assignments along with the total number of class predictions.

Convolutional Neural Networks: A type of artificial neural networks that make use of convolutions, mainly to process images.

F1-Score: Overall performance measure based on a weighted average between precision and recall. F1 is usually more useful than accuracy, especially in the case of class imbalance.

Hyperparameter: All the variables that a user can set before starting the training, which are tunable and can directly affect how well an algorithm learns.

Loss Function: Mathematical function that is used to quantify how good or bad a model is performing.

Precision: The ratio between correctly predicted positive (target) observations to the total amount of observations that were predicted as positive. It is used to determine how often a model is correct when its prediction is positive.

Recall: Also known as sensitivity, it measures how often a model correctly identifies an observation for which both the predicted and the actual class labels are positive.

Visual Inspection: The method of looking for defects using the naked eye and non-specialized inspection instruments.

Chapter 5

The Role of AIoT–Based Automation Systems Using UAVs in Smart Agriculture

Revathi A.
VISTAS, India

Poonguzhali S.
VISTAS, India

ABSTRACT

Agriculture is not all about food and includes production, promotion, filtering, and sales of agricultural products and provides better employment opportunities to many people. Nowadays, precision agriculture is gaining more popularity, and its main goal is the availability to all common people at low cost with maximum crop productivity. It also helps in protecting the environment. IoT, internet of things, a technology that is developing in modern society, can be applied to agriculture. At present, IoT-enabled technology in agriculture has developed to a greater extent, particularly a drastic development of unmanned aerial vehicles (UAVs) and wireless sensor networks (WSN), and could lead to valuable but cost-effective applications for precision agriculture (PA), including crop monitoring with drones and intelligent spraying tests. In this chapter, the authors explore the various applications of artificial intelligence of things (AIoT) and provide detailed explanation on how AIoT may be implemented in agriculture effectively. Moreover, they highlight crucial future research and directions for AIoT.

DOI: 10.4018/978-1-6684-4991-2.ch005

INTRODUCTION

Agriculture, the backbone of Indian economy, plays a dynamic role in the production of essential food crops and providing staple food for more than a century. Researchers are applying a various technique to improve agricultural practices such as crop management, automated vehicle, climate conditions etc. IoT technologies can offer high potential in smart agriculture and precision agriculture. The various IoT devices and technology used in precision agriculture are used for monitoring, analyzing and protecting the environment for increasing productivity and reducing crop damage. But change in climate poses various challenges that can affect many sectors, including agriculture. According to the Food and Agriculture Organization (FAO) of the UN and the International Telecommunication Union, the world's population must find new innovations to rise food production by 60% by 2040. There are too many advances in precise agriculture today to increase crop yields. Especially in developing countries such as India, more than 60% of the rural population rely on agricultural land (L. M. Gladence et al., 2020).

The solution for the important task is the proper implementation and use of information and communication technology services, which provide the opportunity to intensification of the use of agrochemical products such as pesticides and fertilizers while minimizing operating costs. For example, aerial crop monitoring and intelligent spraying. UAV is one of the aircrafts that can fly autonomously without a pilot. UAVs are much easier to operate and cost-effective than manned aircraft. They are also more effective than ground-based systems and have the capability to cover a wide area without any damage in a short time. Mini UAVs, also known as drones, are characterized by being more cost-effective and capable of spraying pesticides. With the evolution of information technology, low-altitude remote sensing technology, represented by the Internet of Things (IoT) and unmanned aerial vehicles (UAV), is being proactively used in the field of environmental monitoring. When modernizing agriculture, IoT and UAVs can track the incidence of crop diseases and pests in terms of ground-based micro and macro-economic indicators, respectively. IoT technology can be used to collect weather parameters of crop growth in real time using a variety of low-cost sensor nodes. Depending on the spectral camera technology, UAVs can take images of farmland, but these images can be used to analyze the presence of pests and diseases in crops. The agricultural sector suffers huge losses from diseases. The diseases are occurring by pests and insects that reduce crop yields. Pesticides and fertilizers are generally sprayed to eliminate insects and pests to improve crop quality. UAV - Airplanes are used to spray pesticides to prevent human health problems when manually spraying. UAVs can easily be used where human intervention is difficult. To reduce the cost and the labor shortage, Computer systems that are capable of doing activities that typically require human intelligence are referred to as having artificial intelligence. Features of human intellect like as speech recognition, decision-making, and visual perception may be possessed by artificial intelligence. For example, Manufacturing robots, self-driving cars, smart assistants etc. Artificial intelligence (AI) in the form of machine learning (ML) enables software applications to improve their propensity to predict outcomes without having to be explicitly programmed. In order to forecast new output values, machine-learning algorithms utilize past data as the input. Machine learning algorithms can be classified into four categories: supervised, semi-supervised, unsupervised, and reinforcement learning. AI, which nowadays can be combined with IOT, symbolizes the age of artificial intelligence of things (AIoT) (S. S. Priya et al., 2020, Bansod et al., 2017).

This study explores AIoT and shows how AI can make IoT devices work smarter, faster and safer. It also briefly introduces AIoT and its architecture in agriculture to predict the crop disease and explain UAV to spray the pesticides and fertilizer. Also discusses applications of AIoT. Finally, the challenges faced by AIoT and potential research opportunities are presented.

BACKGROUND

The term "smart" is used to describe a device when AI and IoT are integrated. IoT devices may learn from gathered data and assess it, generate insights from it, and use those insights without the assistance of a human. When data is processed in the IoT, the AIoT is born. These devices develop "intelligent" and verbal capabilities. AI strongly influences IoT through IoT and machine learning technology. Through networking, signals, and data sharing, AIoT places a high value on AI and is revolutionary and dependent on both kinds of technology. Un - structured data collected from both humans and robots will increase as IoT networks spread throughout big businesses. Data analytics programs can benefit from IoT-generated data with the aid of AIoT (Gao D et al., 2019).

If it comes to AIoT development, there are many stages. The very first stage is something we're all familiar with: connecting two devices and allowing them to be operated via a remote control. The second level entails linking to the cloud in order to give AI inference automatically. In third stage, there is no need for a cloud connection because the gadgets have their own intelligence. Peer-to-peer device communication is the final stage. Devices are "smart," meaning they can talk with one another, share information, and collaborate on tasks. These latter two stages necessitate the use of an AI chip. While the notion of AIoT is indeed relatively new, it offers several opportunities to improve industry verticals such as enterprise, industrial, and consumers' customer relations industries, and these opportunities will only grow as the technology matures. The Internet of Things (IoT) could be a realistic solution to existing operational issues like the cost of good human capital management (HCM) or even the intricacy of distribution networks and service models (Al-Rubaye et al., 2019).

Machine Learning

The backbone of AIoT is machine learning, a form of AI that allows software to predict events without it being programmed explicitly by teaching itself on past data. It is capable of mimicking intelligent human behavior. Machine learning may be used to predict text, translate languages, recognize photos in order to diagnose medical issues, and even power self-driving automobiles. Machine learning models can discover distinguishing characteristics of a system that aren't defined by a theoretical equations. These models can effectively execute tasks such as classification, regression, and arithmetic computations, allowing modelling for accessibility, mobility, and managing IoT device network connections. It also increases network management performance by keeping the current Key Performance Indicator (KPI) within pre-defined thresholds (Souza et al., 2018). The overview of Machine Learning concept is shown in Figure 1.

Supervised Learning

One of the ML methods that employs labelled data sets to train the model is supervised learning. After the input data is submitted to the model, a cross-validation technique is used to alter the weight of the model to suit it well. Supervised Learning is mostly used to classify data or properly forecast outcomes. Linear Regression, Logistic Regression, Nave Bayes, Support Vector Machine, Decision Tree, and Random Forest are some of the most often used supervised learning techniques. An example for Supervised Learning is given in Figure 2.

Unsupervised Learning

Without any human interaction, the unsupervised learning approach takes an unstructured data set and analyzes the functionalities in the data to reflect underlying data structures and patterns. Unsupervised Learning is a great approach for data processing, image identification, and consumer segmentation since it can find similarities and contrasts in data. K-means clustering, Principal Component Analysis (PCA), Apriori algorithm, Hierarchical Clustering, and Neural Networks are some of the most often used approaches in Unsupervised Learning. An example for unsupervised Learning is given in Figure 3.

Figure 1. Overview of Machine Learning

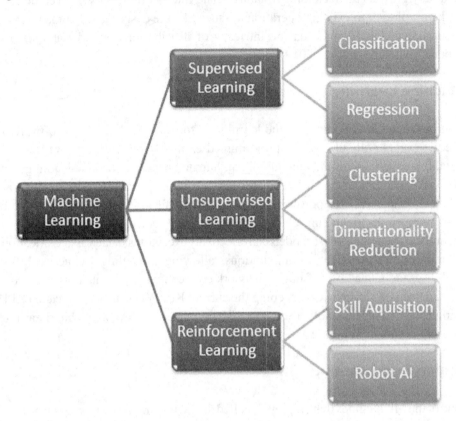

Figure 2. Supervised Learning

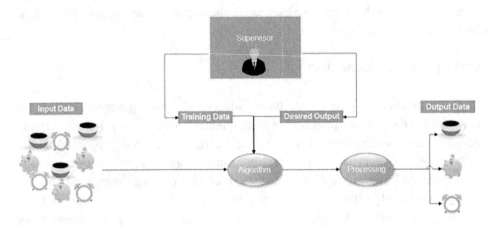

Figure 3. Unsupervised Learning

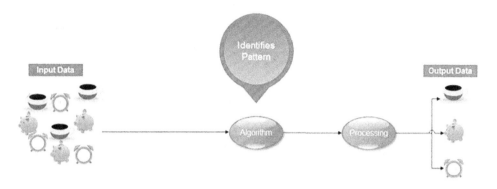

REINFORCEMENT LEARNING

Reinforcement Learning is a merit-based approach to learning. The model prizes the activity if it is a preferred behavior and punishes it if it is an unwanted behavior, and it performs the actions by trial and error. This strategy finds the best answer by focusing on the long-term and highest payoff. State-action-reward-state-action (SARSA), Q-Learning, Markov Decision Process (MDP), and Deep Q-Networks are some of the most often used Reinforcement Learning approaches (Pascuzzi S et al., 2018). An example for Reinforcement Learning is given in Figure 4.

APPLICATIONS OF AIOT

Many AIoT applications are currently focused on the installation of cognitive computing in consumer appliances, and many are retail goods centered. Some of the applications of AIoT are given below.

SMART FACTORIES

The term "smart factory" refers to computerized production, which includes IoT sensors, artificial intelligence, robots, and machine learning in the factory's inner workings. Digitization delivers useful insights that can assist manufacturers improve quality control, machine predictive maintenance, and communication between the information technology (IT) as well as operational technology (OT) sides of the business. Robots and networked automated systems are part of the smart industry. AI and machine learning algorithms can help achieve the goal of a smart factory.

Figure 4. Reinforcement Learning

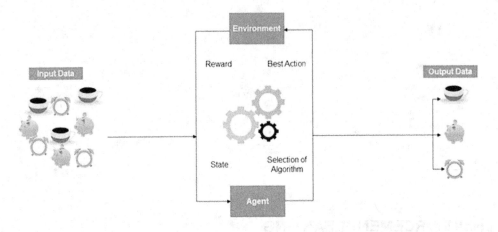

The connectivity of sensors, microchips, actuators, controllers, and robotics can automate activities like asset monitoring, which checks the status of materials and products, and digital twins, which simulates the outcomes throughout an asset's existence. These inspection systems will not only automate the procedure, but will also self-improve by adjusting the process based on past data and analytics (Bah et al., 2017).

SMART PRODUCT

A smart product has intelligence built in to read, adjust, and react to its working environment. Smart products have sensors that allow them to detect their environment, as well as electronics, embedding code, and on-board systems that decode the signals received from the sensors. Sensor data is then either stored locally or sent to and from the cloud over the internet, based on the system design logic. Finally, either locally – on the edge – or in the cloud, machine learning (ML) algorithms use this data to determine the proper action and trigger the actuators that carry it out. Through dashboards and alerts, data analytics also enable real-time visualization to users and product manufacturers. Because microchips and sensors are used in Industry 4.0 items, they are intelligent. The existing manufacturing system must be integrated with the industry 4.0 architecture (IoT+WSN+CPS), which will allow products and humans to communicate. Smart products offer a slew of benefits to both end users and producers. Smarter products can improve end-user experience by delivering control, automation, and flexibility to the fingertips of users, as well as analytical data for enhancing product performance and lowering running costs.

Smart products allow manufacturers to make data-driven product decisions that increase operational efficiency, product quality, and time-to-market, as well as provide prompt assistance. Smart products are more competitive and future-proof, and they provide the possibility of recurring revenue streams. Despite their many advantages, designing linked products necessitates a different mentality. All design decisions must take into account the product's capacity to link and communicate, as well as the limited or complete autonomy that knowledge can provide. The versatility and scalability that comes with an IoT system interconnecting many of such devices, as well as how the product must work in such a linked environment, must also be considered (Pantelej E et al., 2018).

SMART CITIES

IoT technology can help large regions enhance different aspects of its infrastructure and common areas. To be more specific, such technologies allow regional leaders to quickly track and manage things via the internet. Sensors and related devices are the focus of such technologies, which are beneficial all around the world. Indeed, given the wide range of applications and the vast geographic area it can cover, employing the IoT for Smart Cities is a viable option. With its assistance, citizens can be safe and have access to adequate infrastructure. In addition, cities have less pollution and traffic. In 2050, the world's cities are expected to have a population of almost 6 billion people. This fast growth in numbers may result in both Big Data and service demand. Many domains, including as smart health centers, smart agriculture, smart institutions, smart homes, smart offices, and smart transportation, are required to develop the future smart city. In the topic of smart cities, a lot of study has been done.

IoT (internet of things) to collect and analyze data, Smart City makes use of a variety of IoT devices. Meters, lights, and any associated sensors are some of the sources of data for such systems. In context of general utilities, infrastructures, and other services, these technology cities are rapidly evolving. Person in control of the city's operations can use data they collect to identify user demands and issue areas. The city handlers can immediately address their citizens and grasp what they demand with the use of the internet. They can also keep track of what's going on in the area and ensure proper growth (Parraga A et al., 2018).

INTELLIGENT HEALTHCARE

To obtain inexpensive, high-quality healthcare, people must have a better understanding of diseases and their health. The public health sector is currently undergoing a significant transformation that is increasingly focusing on the convergence of various health paths defined not only conceptually but also methodologically through the revamp of services and the diffusion of IoT and sensors. People's health must be monitored for the prevention, well-being, and treatment of difficult chronic diseases, which necessitates the data integration from clinical testing with data from the real world. Automated analysis systems based on machine learning and artificial intelligence achieving objectives versatile interpretation of massive amounts of data. Research in related fields suggests that remote monitoring of health is feasible, but the advantages it might provide in different situations are probably more crucial. By monitoring non-critical patients at home instead of in the hospital, remote health monitoring could free up hospital resources such as doctors' time and beds. It might help rural communities gain improved healthcare access or prolong the independence of senior citizens living at home. In essence, it can broaden access to healthcare services while decreasing pressure on healthcare systems and giving people a greater sense of control over their own health at all times (Mancini A et al., 2018).

SMART AGRICULTURE

Using IoT, big data, and advanced analytics technology, smart agriculture refers to a broad range of agricultural and food production tactics. For people and the globe at large, IoT in agriculture might be a game-changer. As a result of extreme weather, deteriorating soil, drier regions, and collapsing ecosystems, food production is becoming more challenging and expensive. The Internet of Things offers hitherto impossible efficiency, resource and cost savings, digitization, and data-driven processes in agriculture as well as other industries. But, in agriculture, these advantages aren't improvements; rather, they're cures for a sector-wide plethora of grave problems. Agriculture that is data-driven helps in growing ever better goods.

Smart Decision Support Systems (SDSS) are being used in the agricultural sector with the goal of assisting farmers and people interested in agricultural investment in making sound decisions. There are several different decision support systems for crop cultivation, including those for managing irrigation, fertilization, and other service activities.

Agricultural soil sensors, overhead drone monitoring, and farm mapping are tools that help farmers better understand the intricate connections between the environment and the state of their crops. Using interconnected systems, they may duplicate the ideal conditions and increase the nutritional value of the products. Unmanned aerial vehicles, or drones, have become more and more common in the sector. Drones are typically employed in sustainable farming as a surveillance system powered by the Internet of Things, as well as tools for farm planning, on-demand spraying, and pesticide application. An automated plant security drone named the Xaircraft P30 was recognized with a Red Dot award. In order to accomplish remarkable flying abilities and appropriate chemical spraying, it uses sophisticated algorithms, which can save up to 30% of pesticide and 90% of water. On the other side, smaller drones carry out an important activity called pollination. Some businesses have chosen to concentrate on producing microdrones which may perform bee tasks in agriculture as a result of the alarming mass beehive collapses and significant drop in the bee colonies (Palomino W et al., 2018).

Regarding to the discussed methodologies and applications, few of the state-of-the-art works have been provided, C. –j. Chen *et al* (C. -J. Chen et al., 2020) explained that Artificial Intelligence and Image Recognition technologies are used in conjunction with environmental sensors and the Internet of Things (IoT) to identify pests. The Long Short-term Memory (LSTM) algorithm is used for predicting the occurrence of pest. The suggested method alerts farmers to the presence of certain pests before they become a major problem. It increases the overall economic worth of agriculture by offering appropriate pest management strategies that limit crop losses and the environmental damage caused by chemical overuse. The proposed model got an overall accuracy of 92.7%. Panagiotis Radoglou-Grammatikis *et al* (Panagiotis Radoglou-Grammatikis et al., 2020) provided a deep survey on precision agriculture with its technologies and aspects like mapping of soil, mapping of production and the uses of GPS and GIS system. It also gave various analysis on different types of drones (UAVs) along with its architecture, payload and characteristics. It finally concluded that the use of UAVs can enable monitoring as well spraying fertilizers and pesticides which can detect and eliminate the pests respectively. Gao D *et al* (Gao D et al., 2020) proposed a framework to deal with the pests in agricultural land which is said to be related with the weather. The frame was divided into 3 parts. The first part is the use of sun trackers to enable energy harvesting for drones. The second part is the use of flight mode in UAVs to maximize the flight time of the drones. Finally, the images which are captured by the drones, are transferred to the cloud for storage where spectrum technology is used to analyze the damage done by pests. It finally concluded that agricultural pest and disease prevention and control is a systematic effort that poses a considerable challenge for farmers and researchers and it necessitates long-term monitoring and analysis employing information technology. Erwin Kristen *et al* (Erwin Kristen et al., 2021) discusses some of the cyber vulnerabilities that have been discovered, as well as some first ideas for how to resolve them. The author provided a brief introduction, research output and the architecture of the European project called "AFarCloud". It showed the various types of cyber-attacks involved in the farm management and recommended some software assets as well as communication protocols to improve the cyber security in the farm management. It finally concluded that there is a necessary to spread awareness on cyber-security guidelines for modern agriculture technologies to overcome various vulnerabilities. Jing Zhang *et al* (Jing Zhang et al., 2021) provided a detailed survey on AIoT and the AI methods which can improve the IoT devices to make the devices smarter, faster and safer. It also presented various application as well as challenges faced by the AIoT. It finally concluded that the three tier architecture of AIoT has provided various resources which can be found useful for the deep learning for lightweight models. UM Rao Mogili *et al* (UM Rao Mogili et al., 2018) provided a brief survey on how the drones can be implemented in the agricultural land for crop monitoring and spraying of pesticides. It has presented various types of UAVs as well as its applications which can be found useful for the implementation of UAVs in various fields. It finally concluded that the precision agriculture has improved greatly in the past decade with latest technologies like drones to enable remote monitoring system as well as automatic spraying of pesticides and fertilizers.

AIOT IN SMART AGRICULTURE

The Internet of Things (IoT) has some key components, including sensor-equipped devices, wireless communication technologies, internet access, sensed and transferred data, etc. The efficient implementation of IoT systems, which can be divided into categories based on spectrum, transmission range, and application scenarios, depends heavily on the wireless communication technology. The three layers that make up the core of the Internet of Things' structure are the perception layer, where sensing takes place, the network layer, which handles data transport, and the application layer, which handles data storage and manipulation (Villa-Henriksen A et al., 2020).

1. **Perception layer:** Various terminal devices, sensors, Wireless Networks (WSN), RFID tags and reader devices, and other items make up the perception layer (Tzounis A et al., 2017). Sensors are employed at this stratum to gather weather data, air velocity, dampness, nutrients level, pest infestations, predatory insects, and other variables. The information is analyzed by embedded devices and uploaded to a higher layer for additional processing and analysis through the network layer. Agricultural and livestock products are tracked, monitored, controlled, and identified using these terminal devices and sensors.

2. **Network layer:** In this article, IoT's network layer is concentrated, where sensors and devices must be connected to other nodes nearby and gateways in order to form a network. In order to transfer data to a remote infrastructure, where it will be stored, further analyzed, processed, and distributed to valuable information, the sensor nodes connect and communicate with other nodes and gateways inside a network at this layer. There is a substantial body of

scientific research on wireless networks that addresses a number of issues, such as lowering energy consumption, enhancing networking capabilities, and boosting efficiency and scalability (Tzounis A et al., 2017); however, the connectivity issues in remote areas have not received as much attention. The most dependable long-range IoT connectivity technologies are IEEE 802.11ah and LoRa/LoRaWAN. The first is a modification to the IEEE 802.11 family that was launched in 2017 to assist IoT use cases like smart metering (Bacco, M et al., 2018). In comparison to Bluetooth and IEEE 802.15.4, it employs 900 MHz license-exempt channels, offers a broader coverage area, and requires less energy. With a single connection point, it offers connection to numerous devices within a one kilometre coverage radius. However, among the most intriguing LPWAN standards for a networking of rechargeable batteries wireless nodes is LoRaWAN.

3. The application layer is the top layer of the IoT architecture, and it is here that the advantages and uses of IoT are most readily visible. This layer has many sophisticated platforms or systems for monitoring and managing the health of the soil, the availability of water and nutrients, plants, and animals. As an outcome, output efficiency can be increased. These layers also enable the early detection and management of illnesses and insect pests, infestation, and agricultural product safety tractability.

UAV'S IN SMART AGRICULTURE

Agricultural robots are one of the most practical developments in smart farming, and UAV, sometimes known as drones, have been widely used (Muchiri N et al., 2016). Farmers frequently employ drones or UAVs for regulating and monitoring farm growth. Some UAVs are used to effectively water the plant and other pesticides in difficult terrain where human movement is difficult and the crops have varying heights. Groups of drones outfitted with diverse sensors and 3D cameras can cooperate thanks to recent developments in swarm technology and task control to give farmers complete land management tools. By drastically decreasing working hours, these agricultural UAVs let farmers to have a bird's eye perspective of their fields, allowing for better management and control of the farms. This increases stability, production, and measurement accuracy. Additionally, their applications have helped many aspects of agriculture, including the prospecting and application of fertilizers and pesticides, the identification and elimination of weeds, the planting of seeds, the evaluation of soil fertility, mapping, etc.

Despite UAV developments, a few problems remain that must be addressed for better deployment. These problems include battery efficiency, short flight times, long communication distances, and payload. For instance, because energy is a limited resource for UAVs, researchers have concentrated on lowering energy usage in UAVs (Islam N et al., 2019). Addressing the connectivity issue, however, still needs additional focus. So that academics can conduct research to address these problems, researchers are outlining the connection restrictions of agricultural UAVs in this work. UAVs can be used in a variety of smart farming application areas.

1. Monitoring is the timely detection of several farm parameters. One of the initial components of smart agriculture is automatic monitoring. Strategically positioned sensors can automatically detect and send data to a gateway for additional processing and analysis. Crop metrics like the index of leaf area, plant height, and the color, shape, and size of the leaves (Dadshani S et al., 2015) are all monitored by sensors. Additionally, they can be used to monitor soil moisture, irrigation water parameters like salinity and pH level, and weather factors including air density, ph, humidity levels, air velocity, wind speed, rainfall, radiation, and more. Additionally, remote sensing is employed quite efficiently.

2. Effective 2D or 3D maps about an agricultural land can be produced using UAVs. These maps can be used to detect things like the size of the field, the kind of soil, the quality of the crops, and crop infestation. For example, the authors in (Muchiri N et al., 2016) used UAV photos to produce high-resolution maps outlining the regional variations of radiation interception. These maps are utilized for lucrative precision farming jobs including separating fruit quality areas, locating areas of deforestation, and controlling homogeneous zones agronomically.

3. The use of UAVs for weed and pest infestation identification is another crucial aspect of smart farming. According to a US research, crop diseases and insect infestations cost the US economy about $40 billion annually in losses. Therefore, to minimize this damage, early diagnosis is crucial. Researchers from (Turner D et al., 2011) carried out a study to evaluate the vegetation indices of grapes by collecting information from vineyards using multispectral cameras placed on a UAV. These observations can be applied to a variety of tasks, including weed mapping, pest infestation identification, and weed detection. To find weeds and pests in plants and crops, many researchers have put RGB cameras, hyper spectrum cameras, and multi-spectrum sensors on unmanned aerial vehicles (UAVs).

4. Undoubtedly, deploying UAVs will increase the efficiency of seed and seedling planting. For instance, (Salaan C.J et al., 2019) the authors demonstrate the efficient use of UAVs over a sizable region of uneven rice paddies. For efficient and prompt distribution of seeds, fertilizer, and plant nutrients, they used a

UAV-based system. The use of UAVs to plant seeds and seedlings is still in its infancy; researchers are working to create UAVs with image recognition technology and an improved planting method.

5.   In comparison to a fast spraying or a wide-area sprayer, UAVs have demonstrated their ability to spray pesticides and fertilizers efficiently and quickly. Risks of worker illnesses and environmental contamination are correlated with the amount of pesticides used per hectare of agricultural. UAVs enable large-scale cleaning of up to 50 hectares every day and require just around 10 minutes of labor per 0.5 hectare area, reducing the use of pesticides. Utilizing UAVs to replace labor is one of the main goals. Using a UAV to spray fertilizer from different heights, for instance, writers in (Pan Z et al., 2016) carried out research on citrus farms to ascertain the ideal level of preventive labor.

OPEN RESEARCH ISSUES

Hardware management and scarce energy capacity: The perception layer's hardware is put up in difficult environments like farms and mines, which have high temperatures, rain, strong winds, and excessive humidity, among other extreme weather conditions. As a result, the physical devices' electronic circuits are harmed. Therefore, it is necessary to develop hardware that is more durable and less susceptible to damage from the environment. Additionally, these devices run continuously for a long time on insufficient battery power. Alternative energy-efficient solutions are therefore necessary for the edge devices because fast battery replacement in the event of a program failure is challenging, especially in remote places.

Issues with security and privacy: The growing usage of IoT increases susceptibility to cybersecurity threats and weaknesses in smart farming. Authentication process and confidence, data, network and adherence, and supply - chain management were some of the major issues for privacy and security in smart farming that Gupta et al. introduced in (Gupta M et al., 2020). The design of smart farming takes into account the likelihood of cyberattacks, which is a concern. a closely interconnected system called a "smart farm" that produces a ton of data Since most of the equipment for use in smart farming is unsupervised, an attacker can readily target it.

Big data in smart agriculture: IoT and UAV-based smart farming captures, stores, and analyses a vast amount of information with a broad range for decision-making. By offering real-time operational decisions, big data are leveraged to deliver actionable analytics in farming operations. Beyond core production, Big Data uses in Smart Farming have an impact on the whole food supply chain. Quality of the data, smart analysis and processing long-term integration of Big Data sources, etc., are the main problems with data analysis. The platform's transparency is also crucial because it can give farmers more control over their place in supply chains.

Automated watering management and control in rural areas: Automated watering management and control in rural areas is crucial to upholding the aesthetic standards of parks and sporting venues. According to the needs of the crops, farms must water. Additionally, effective water management must be put in place to stop an excess quantity of water from penetrating the soil's lower layer and leaking nutrients into the stream. Most of the aforementioned goals can be met by using live moisture monitoring at various soil depths, AI-based watering, automated watering, and management of water resources using IoT and UAV technology.

SOLUTIONS AND RECOMMENDATIONS

This study examines a range of applications for smart farming, the advantages and applications of combining IoT and UAVs in farming, various communication systems, and the challenges and limitations of IoT and UAV connections in remote locations. The authors have looked at the specific case of using IoT for smart agriculture in remote areas in addition to the connection constraints with regard to communications technologies and transmission range. The paper provides a summary of the uses of smart farming and helps academics recognize issues and open-ended research questions that are important for identifying possibilities and limitations for smart and sustainable farming.

CONCLUSION

This research assessed a range of applications for smart farming, the advantages and applications of combining IoT and UAVs in farming, various communication systems, and the challenges and limitations of IoT and UAV connectivity in remote locations. This helps the farmers to prevent the crop from the disease and spray pesticides automatically without human intervention. This helps to improve the crop yield and quantity as well the quality of the crop. The implementation of AIoT in agriculture can bring forth many benefits such as automatic disease prediction, crop management, monitoring the field etc. AIoT technology helps farmers and protect the agriculture environment as well as improves the Indian economy.

REFERENCES

Ali, A. H., Chisab, R. F., & Mnati, M. J. (2019). A smart monitoring and controlling for agricultural pumps using LoRa IOT technology. *Indonesian Journal of Electrical Engineering and Computer Science*, *13*(1), 286. doi:10.11591/ijeecs.v13.i1.pp286-292

Asnafi, M., & Dastgheibifard, S. (2018, July 9). A Review on Potential Applications of Unmanned Aerial Vehicle for Construction Industry. *Sustainable Structures and Materials. International Journal (Toronto, Ont.)*, *1*(2), 44–53.

Bah, M. D., Hafiane, A., & Canals, R. (2017). Weeds detection in UAV imagery using SLIC and the hough transform. *2017 Seventh International Conference on Image Processing Theory, Tools and Applications (IPTA)*. 10.1109/IPTA.2017.8310102

Bansod, B., Singh, R., Thakur, R., & Singhal, G. (2017). A comparision between satellite based and drone based remote sensing technology to achieve sustainable development: A review. *Journal of Agriculture and Environment for International Development*, *111*(2), 383–407. doi:10.12895/jaeid.20172.690

Chen, C. J., Huang, Y. Y., Li, Y. S., Chang, C. Y., & Huang, Y. M. (2020). An AIoT Based Smart Agricultural System for Pests Detection. *IEEE Access: Practical Innovations, Open Solutions*, *8*, 180750–180761. doi:10.1109/ACCESS.2020.3024891

Dadshani, S., Kurakin, A., Amanov, S., Hein, B., Rongen, H., Cranstone, S., Blievernicht, U., Menzel, E., Léon, J., Klein, N., & Ballvora, A. (2015). Non-invasive assessment of leaf water status using a dual-mode microwave resonator. *Plant Methods*, *11*(1), 8. doi:10.118613007-015-0054-x PMID:25918549

Gajja, M. (2020). Brain Tumor Detection Using Mask R-CNN. *Journal of Advanced Research in Dynamical and Control Systems*, *12*(SP8), 101–108. doi:10.5373/JARDCS/V12SP8/20202506

Gao, D., Sun, Q., Hu, B., & Zhang, S. (2020). A Framework for Agricultural Pest and Disease Monitoring Based on Internet-of-Things and Unmanned Aerial Vehicles. *Sensors (Basel)*, *20*(5), 1487. doi:10.339020051487 PMID:32182732

Gao, D., Zhang, S., Zhang, F., He, T., & Zhang, J. (2019). RowBee: A Routing Protocol Based on Cross-Technology Communication for Energy-Harvesting Wireless Sensor Networks. *IEEE Access: Practical Innovations, Open Solutions*, *7*, 40663–40673. doi:10.1109/ACCESS.2019.2902902

Gladence, L. M., Anu, V. M., Rathna, R., & Brumancia, E. (2020). Recommender system for home automation using IoT and artificial intelligence. *Journal of Ambient Intelligence and Humanized Computing*. Advance online publication. doi:10.100712652-020-01968-2

Gupta, M., Abdelsalam, M., Khorsandroo, S., & Mittal, S. (2020). Security and Privacy in Smart Farming: Challenges and Opportunities. *IEEE Access: Practical Innovations, Open Solutions*, *8*, 34564–34584. doi:10.1109/ACCESS.2020.2975142

Islam, S., Sithamparanathan, K., Chavez, K. G., Scott, J., & Eltom, H. (2019). Energy efficient and delay aware ternary-state transceivers for aerial base stations. *Digital Communications and Networks*, *5*(1), 40–50. doi:10.1016/j.dcan.2018.10.007

Kristen, E., Kloibhofer, R., Díaz, V. H., & Castillejo, P. (2021). Security Assessment of Agriculture IoT (AIoT) Applications. *Applied Sciences (Basel, Switzerland)*, *11*(13), 5841. doi:10.3390/app11135841

Mancini, A., Frontoni, E., & Zingaretti, P. (2018). Improving Variable Rate Treatments by Integrating Aerial and Ground Remotely Sensed Data. *2018 International Conference on Unmanned Aircraft Systems (ICUAS)*. 10.1109/ICUAS.2018.8453327

Mogili, U. R., & Deepak, B. B. V. L. (2018). Review on Application of Drone Systems in Precision Agriculture. *Procedia Computer Science*, *133*, 502–509. doi:10.1016/j.procs.2018.07.063

Palomino, W., Morales, G., Huaman, S., & Telles, J. (2018). PETEFA: Geographic Information System for Precision Agriculture. *2018 IEEE XXV International Conference on Electronics, Electrical Engineering and Computing (INTERCON)*. 10.1109/INTERCON.2018.8526414

Pan, Z., Lie, D., Qiang, L., Shaolan, H., Shilai, Y., Yan-de, L., Yongxu, Y., & Haiyang, P. (2016). Effects of citrus tree-shape and spraying height of small unmanned aerial vehicle on droplet distribution. *International Journal of Agricultural and Biological Engineering*, *9*, 45–52.

Pantelej, E., Gusev, N., Voshchuk, G., & Zhelonkin, A. (2018). Automated Field Monitoring by a Group of Light Aircraft-Type UAVs. *Advances in Intelligent Systems and Computing*, 350–358. doi:10.1007/978-3-030-01821-4_37

Parraga, A., Doering, D., Atkinson, J. G., Bertani, T., de Oliveira Andrades Filho, C., de Souza, M. R. Q., Ruschel, R., & Susin, A. A. (2018). Wheat Plots Segmentation for Experimental Agricultural Field from Visible and Multispectral UAV Imaging. *Advances in Intelligent Systems and Computing*, 388–399. doi:10.1007/978-3-030-01054-6_28

Pascuzzi, S., Anifantis, A. S., Cimino, V., & Santoro, F. (2018). *Unmanned aerial vehicle used for remote sensing on an Apulian farm in Southern Italy*. Engineering for Rural Development. doi:10.22616/ERDev2018.17.N175

Radoglou-Grammatikis, P., Sarigiannidis, P., Lagkas, T., & Moscholios, I. (2020). A compilation of UAV applications for precision agriculture. *Computer Networks*, *172*, 107148. doi:10.1016/j.comnet.2020.107148

Salaan, C. J., Tadakuma, K., Okada, Y., Sakai, Y., Ohno, K., & Tadokoro, S. (2019). Development and Experimental Validation of Aerial Vehicle with Passive Rotating Shell on Each Rotor. *IEEE Robotics and Automation Letters, 4*(3), 2568-2575. doi:10.1109/LRA.2019.2894903

Souza, I. R., Escarpinati, M. C., & Abdala, D. D. (2018). A curve completion algorithm for agricultural planning. *Proceedings of the 33rd Annual ACM Symposium on Applied Computing*. 10.1145/3167132.3167158

Turner, D., Lucieer, A., & Watson, C.S. (2011). *Development of an Unmanned Aerial Vehicle (UAV) for hyper-resolution vineyard mapping based on visible, multispectral and thermal imagery*. Academic Press.

Tzounis, A., Katsoulas, N., Bartzanas, T., & Kittas, C. (2017). Internet of Things in agriculture, recent advances and future challenges. *Biosystems Engineering*, *164*, 31–48. doi:10.1016/j.biosystemseng.2017.09.007

Villa-Henriksen, A., Edwards, G. T., Pesonen, L. A., Green, O., & Sørensen, C. A. G. (2020). Internet of Things in arable farming: Implementation, applications, challenges and potential. *Biosystems Engineering*, *191*, 60–84. doi:10.1016/j.biosystemseng.2019.12.013

Zhang, J., & Tao, D. (2021). Empowering Things With Intelligence: A Survey of the Progress, Challenges, and Opportunities in Artificial Intelligence of Things. *IEEE Internet of Things Journal*, *8*(10), 7789–7817. doi:10.1109/JIOT.2020.3039359

Chapter 6
A Study on the Use of IoT in Agriculture to Implement Smart Farming

Indu Malik
Gautam Buddha University, India

Anurag Singh Baghel
Gautam Buddha University, India

ABSTRACT

Presently there is a massive enhancement in technologies, and a lot of things, appliances, and techniques are accessible in the agriculture sector. One of the famous techniques is known as IoT. Several applications of IoT are evident in the field of agriculture for the benefit of the farmers and in turn for the successful development of the nation. IoT is used in agriculture to improve productivity, efficiency, and the global market. It also helps farmers to reduce manpower, cost, and time. In this chapter, the authors are discussing IoT with the cloud for enhancing smart farming. Smart farming is a concept that is focused on providing the agricultural industry with the infrastructure to leverage advanced technology including big data, the cloud, and the internet of things (IoT) for tracking, monitoring, automating, and analyzing operations. IoT with the cloud is used in various fields of agriculture to improve time efficiency, water management, crop monitoring, and land management. It protects the crops from pests and is also used to control pesticides in agriculture.

DOI: 10.4018/978-1-6684-4991-2.ch006

INTRODUCTION

The current and future era of smart computing rely enormously on the application Internet of Things (IoT) in various areas. In current scenario, IoT is playing a vital role to transform Traditional Technology from normal life to official life everywhere you will find computer. IoT is getting importance in research across the world however it is popular for wireless communications. Basically IoT is a term, which is use for finding uniquely identical objects, things and their respective virtual representations in the internet. Kevin Ashton is known as father of IoT. Initially IoT was used for chain management, after 2008, it scope has enhanced and its development being used in various products such as smart living, education, medical, organization etc. In current era, it is being used for business (like smart education, agriculture, health care), manufacturing, smart devices and machines.

One of the head areas where IoT research is ongoing and lots of products have lunched, various new products will be lunched to make the activities process smarter, easiest and effective towards for better production is "Agriculture". Agriculture (Ali Hasnain, 2021) sector is one of the sectors that is ensure food and food security for livings. If taking Indian Farmer, who are facing lots of trouble due to farm size, transportation issues, shortage of technology, trade, climate, pest etc. No doubt, ICT work on lots of problem for providing its solution to farmers (Chettri & Bera, 2020), infect ICT based techniques have resolved some problems however that are not enough for efficient production. Now ICT has migrated with IoT, and this migration is known as "Ubiquitous Computing". Agriculture production based on lots of different activities such as soil and plants monitoring, water management mentoring, moisture and temperature monitoring, transportation supply management, infrastructure management, animal and pest control monitoring.

IoT based agriculture is a convergence technology, which is used to create high volume in terms of production quantity, and quality, simultaneously it reduced production cost and manpower. Applying IoT in agriculture field data is collected with the help of GPS and sensors (Dinh et al., 2017, García et al., 2020) (smart sensors), and it is processed by the integration techniques of smart farming along with Big data. With the help of smart farming (Islam et al., 2021) Farmers are capable to improve crop yields as well as make it effective (Anjali et al., 2018). As par the current scenario of agriculture that is surrounded by lots of issues, it must be resolved by IoT based smart farming. To build smart farming in real world, requirement is to develop time to time IoT based products on very fast face.

IoT have being use in different domain of agriculture to enhance smart farming. It improves time efficiency, water management, crop monitoring, and soil management. It is also used to control in agriculture. Using IoT, human efforts get reduced mostly work gets done by smart machines. In bygone days farmers used to level the ripeness

Figure 1. Smart Agriculture Is Combination of Iot And Agriculture

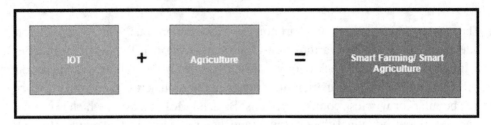

of soil, water, and guided suspicions to develop which to somewhat yield. At that time, they were not able to predict humidity, level of water. And most important they did not predict the climate. Indirectly, Climate plays a vital role in crop production. One of the factors on which the crop yield production is dependent that is climate. If farmer climate prediction is exact that time farmer gets good crop production otherwise it creates terrible for the farmer. Generally, farmers do prediction manually without any smart device because of that, farmers did not get much production as they thought. To get exact information about climate prediction, ripeness of soil, humidity, and level of water using IoT (Internet of thing). Using IoT with agriculture creates a new architecture of agriculture, which is called smart farming. IoT is remodeling of the agriculture business (agribusiness) empower, Farmers, get an extensive range of strategies such as accuracy, data, practical farming to deal with challenges in the field. IoT is used for information assemble according to the circumstances like climate, temperature, water level, soil ripeness, soil PH level. Using IoT plant diseases is detected in early stage due to early detection pesticides spray is also reduced. IoT gives remote access to Farmers. IoT utilizes farmers to get related with his residence from wheresoever and at any point in time. IoT reduces the cost and updates the standard of the developing product. Smart farming or Smart agriculture (Islam et al., 2021) is a combination of two words IoT and agriculture as shown in FIGURE 1.

INTERNET OF THINGS

IoT stands for the Internet of Things and it is a collected device system, which is used to transfer data. The purpose of IoT, is data collection from different resource and share it across the world using wireless internet. Wireless is a network that is use to connect every devices in a virtual model across the internet for data sharing. Before IoT, data collection, data transfer, data storage and communication was very difficult but now a days it becomes easy because of IoT. The current era is based on the technologies and in this era, machines are smart enough, they can communicate with each other (one machine can communicate with another one machine) without any human interaction. A machine can be made smart using technologies. For Example: RFID is a technology, which is used in IoT (Malik & Tarar, 2021). RFID stands for Radio Frequency Identification. RFID is a smart technology, which is used to identify individual objects within machines or computers. RFID technology is also capable to record metadata (metadata is data about data). RFID technology (Malik & Tarar, 2021) core is radio waves, which is used to control a target. It is also used for collecting real-time data. IoT uses RFID technology for object observation with infrared sensors, GPS, and laser scanner sensor equipment. It is capable to capture all the movement of the object with the location. IoT has a vital role to exchange data or information. IoT is an intelligent system with smart devices and sensors. With the help of IoT identification of any object, identification of intelligence, location-tracking, and monitoring all have become easy. IoT devices have been used to collect large amounts of data, including environment, weather, traffic, people, and more. All these elements of data acquisition depend on the accuracy of the analytical capabilities of this technology. The range of IoT sensors is not limited. Basically, it doesn't matter where and how you do it, depending on your needs. Sensors are used in almost every field, including: Sensors used in homes, organizations and agriculture. For example, sensors are used in agriculture to help farmers. Sensors are used in agriculture to determine plant effectiveness and fertilization. Smart farming is a combination of technologies. Smart farming technologies (Mekala & Viswanathan, 2017) include sensors, telecommunications, data analytics, and satellites. Smart farming increases crop yields. In agriculture, various sensors are used to obtain data. Sensors are used for soil scans, water levels, light, crop damage, soil pH and temperature. Telecommunications technologies include networks, GPS, hardware and software. Both hardware and software play an important role in smart farming. Use dedicated hardware and software with IoT sensors to improve crop production. Data analysis is the most important part of smart farming. Used for decision making and forecasting. Data collection is a central part of smart farming (Jayaraman et al., 2016), and the data represent the quality of the yield. Data collection in smart agriculture is related to meteorological data, fertilizers, soil, water and more. FIGURE 2 shown IoT application with devices.

Sensors, satellites and drones are used to collect data. This data is transferred to the AI application for further processing. The combination of these technologies makes it possible to retrieve data between machines. This data is stored in a decision support system so farmers can see what happened at a more basic and detailed level than ever before. It is used here accurately to measure field (Ryu et al., 2022) variability and adjust the required strategy accordingly. This data is used to effectively apply pesticides and fertilizers to crops.

Figure 2. Different Iot Applications

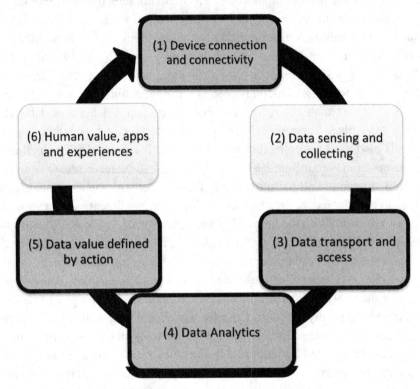

1. **Device connection** Each and every IoT device should be connected with, and it comes under the device connection. When physical devices are connected together to share information through the Internet, it is known as connected devices. These devices are embedded in technologies such as processing chips, software, and hardware. Connected devices are a combination of various hardware such as sensors, computers and mobile phones. These devices can be controlled remotely via a smartphone, tablet, or computer system. These devices are remotely monitored and controlled using smart devices.

2. **Data Sensing** For data sensing, sensors are used, it are physical devices, which have the capacity to sense and capture it as a data. A sensor is an electronic machine, module, or device or subsystem that sends mutual signals to a server to detect, recognize, and further process events around the sensor. Sensors are used to capture events in the environment to collect data. Capturing an event with a sensor is called a data capture. This is also known as sensor data acquisition.

3. **Device Communication** It is a way to transfer data from one node to another across the internet. It does not depend on the location. In computer data is transfer

in electronic waves. IoT device communication is an infrastructure or system for exchanging data or information between devices connected to a system. All devices connected to the system can share data with each other. This system must be connected to the internet to communicate. The data collected by IoT devices is shared with devices connected via the Internet. Sensor data can be sent to the cloud globally or locally for analysis. These devices communicate with other connected devices to get information and data.

4. **Data analytics** A small unit of data processing is easy however a large amount data process using algorithms such as Machine Algorithm. Data analytics is included substantial data analysis, AI (Jha et al., 2019) and cognitive, and the analysis at the edge. Data analysis is used to investigate datasets and draw trends based on trends to draw conclusions. Data analysis is used to retrieve the information contained in a dataset. Specific hardware and software are used for data analysis.

5. **Data value** Data value is a way and it is used to take action on the data. Its perform analysis action, APIs and processes, and actionable intelligence also. The content used to enter records is called a data value. A database is a collection of various data fields such as numbers, names, and contacts. There are different types of IoT data, such as state data and automated data. Status data is structural data, line-based data, used to convey the status of connected devices. Automation data is collected from automated devices such as Automatic lighting was created.

6. **Human value** Human works with smart application. Smart application is not required any human interaction to operate smart applications. Technology has become a part of human life, and the invention of the IoT has made human life easier. Just as human life has grown with technological capabilities, the IoT is growing day by day. With the help of IoT communications, exchanging data and interacting with devices has become easier. IoT works with security. Today, humans move online with IoT devices, not to exchange information. Technology becomes a part of human life, and the more inventions of technology make it easier.

SMART FARMING / SMART AGRICULTURE USING IOT

Smart farming is a concept of latest farming that is a collection of latest technology such as IoT, GPS, Sensors, Big data, cloud data, ML (Machine Learning) (K & S, 2019, "Machine Learning Prediction Analysis Using IoT for Smart Farming," 2020a), and AI (Artificial Intelligence) (Nirav Rathod, 2020) to increase the quality and the quantity of the agriculture production while it optimizes manual efforts and the cost. A technology known as "smart farming" uses information technology to gather data for agricultural management. Smart farming is a system that uses the most recent technology to manage agricultural rather of doing so manually. The idea of autonomous farming has been introduced into smart farming. Agriculture is managed by farmers using technology. Numerous clever techniques and instruments have been added to this breakthrough to improve agricultural production's yields and sustainability. Smart agriculture technology has eliminated the necessity for manual environmental management on the part of farmers. Data from the program can be easily seen and investigated, providing previously unheard-of insight into plant health. IoT technology improves the capacity to predict production outcomes, which helps farmers better plan and distribute their produce. This helps decrease production risk in the agricultural industry. Farmers may reduce labor and waste by using data on the precise batches of crops and the number of crops to harvest, for instance. FIGURE 3 illustrates the core principles of smart farming.

Smart agriculture is technology-based agriculture and focuses on industrial agriculture. Smart Farming insists that farmers use smart farming tools. Smart farming (Miss. Bhagyashree A. Tapakire, 2019) includes the best infrastructure to take advantage of advanced technologies such as big data, cloud and IoT. Agriculture uses more technology to track, monitor, automate and analyze farm operations. Smart farming is created with the help of sensors and software. Thanks to the world's population, smart agriculture is getting better every day. As the population of each country is increasing, the demand for crops is increasing.

IoT has been contributing in each and every area of Industry. In agriculture, IoT is just not providing the solution of the problems, in reality it has changed farming concept with latest technology, which enhanced agriculture production in terms of quality and quantity also.

WHAT IS SMART FARMING?

Smart farming refers technology, which work on real time information and communication to increase agriculture production quality and the quantity. Numerous clever tools and techniques have been provided for this invention to improve yields and the sustainability of agricultural production.

These Technologies are

1. **Sensors:** The sensor is an electronic device (remember that some sensors are not electronic), it takes input from the environment and converts it into data for further processing. This data is further processed either by the machine or human. For smart homes basically, we are using the electronic sensor. Electronic

Figure 3. The Fundamental Elements Needed for Smart Farming

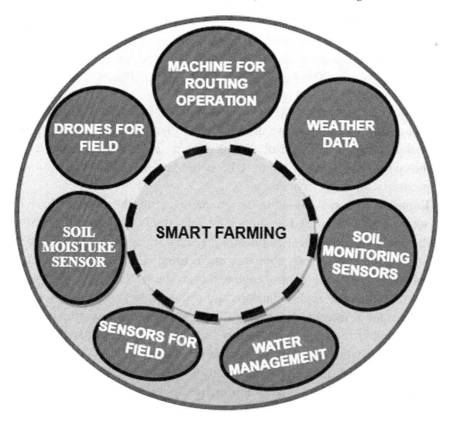

sensors are further categorized into two classes. The first one is an Analog sensor and the second one is a digital sensor. Analog sensors are used to convert sensor data into analog signals whereas digital sensors are used to convert what they are measuring in the input data to digital signals. To create a smart home both analog and digital sensors are used. Sensors could be coupled with any home appliance, such as an air conditioner, gas, light, doorbell, door lock, or any environmental device. With the assistance of sensor data, nowadays it is easy to monitor and control home functionality remotely by the Human.

2. **Software:** Software is a collection of instructions, which are a set of rules for doing work step by step. The computer uses instructions to perform each task. A collection of instructions is called software. The software is used to perform specific tasks. The software is application specific. No single software is used for multiple different applications. The software is in contact with the hardware, and the hardware is operated through the software.

3. **Connectivity:** Connections are used to connect systems or applications. Connection is a way to connect multiple applications or devices together. Connectivity is a way to connect your IoT device to a server or cloud. Data can be transferred from one device to another over the connection. Internet connection is a good example of a connection. If the two devices are not connected, they will not be able to share data.

4. **Location:** The location is the exact location of the person or object where the event occurred. Sensors are used to create intelligent applications. Get the position using the GPS sensors (Malik & Tarar, 2021). GPS is an abbreviation for Global Positioning System. GPS is the best sensor for getting accurate position with smart technology.

5. **Robotics:** Robotics is used to design intelligent machines that help to work automatically. It is used to support people. Robotics is an integrated field that includes mechanical engineering, electrical engineering, mechatronics, information engineering, computer engineering (K. S., 2018), mathematics, software engineering, control engineering and more. Robotics is used to create machines that can replace humans and convey human behavior. Robots can be used for different purposes in different situations, such as crop inspection. Many robots are used in dangerous situations such as radioactive material inspection, explosives detection, and bomb detection.

6. **Data Analytics:** Processing small units of data is easy, but large amounts of data are processed using algorithms such as machine algorithms. Data analysis includes extensive data analysis, AI and cognition, and edge analysis. Data analysis is used to examine datasets, draw trends based on trends, and draw conclusions. Data analysis is used to retrieve the information contained in a dataset. Specific hardware and software are used for data analysis.

CLOUD COMPUTING AND ITS CONTRIBUTION TO IOT FOR SMART FARMING

Cloud computing is a very important and useful technology to share resources and data over the internet in a virtual way. Cloud computing is a virtual service system to provides required resources and data through the internet. Cloud service (Prathibha et al., 2017) could be accessed anywhere at any time without any time and location restrictions. The smart farming is a collection of merged technologies such as IoT, cloud computing, and AI. Instead of this, these technologies are responsible to manage local and central computing with the optimized resource.

A computing task should be executed either on the IoT and smart farming devices or outsourced to the cloud. Where to compute depends on the overhead tradeoffs, data

Figure 4. Different Role of Iot in Agriculture Sectore

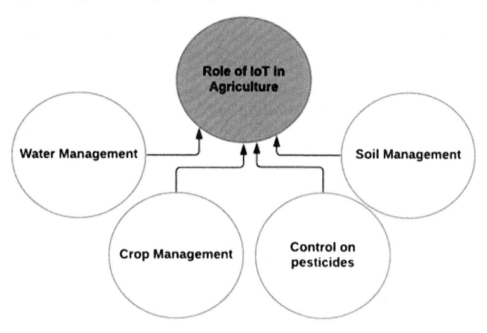

availability, and data dependency, amount of data transportation, communications dependency, and security considerations. On the one hand, the triple computing model involving the cloud, IoT, and smart farming (Kasar, 2019), should minimize the entire system cost, usually with more focus on reducing resource consumption at agriculture. On the other hand, an IoT and smart farming computing service model should improve IoT users fulfill their demands when using cloud applications and address complex problems arising from the new IoT, smart farming, and cloud service model. IoT uses robots, drones, computer imaging, and remote sensors with machine learning to capture continuously processing data in agriculture. In agriculture, to monitor crops (L. S. & B., 2018, Harendra et al., 2020)) IoT uses analytical tools. Role of IoT is represented in FIGURE 4. IoT devices are using in agriculture to surveying and mapping the crops (field) to collect data then provide this data to the farmer for rational farm and management plans. After accessing this data farmers save both time and money.

IoT devices are using in farming to target conventional farming operations (Khoa et al., 2019). IoT helps to increase crops product and decreasing crop production loss. IoT-based remote sensing is availing oneself of sensors that are placed with the field for instance weather stations to gather data which is transmitted to the analytical tools for further analysis. Sensors are devices that are sensitive to anomalies. Farmers can use sensors to monitor the crops and take steps based on insights.

1. **Crop Monitoring** Sensors are placed along with farms by the farmer to monitor the crops change like humidity, temperature (Nirav Rathod, 2020), crop size, and for any disease. If sensors capture an anomaly, then they notify farmers regarding that. Using remote sensing techniques farmers get helps to save their crops from spread diseases.
2. **Weather Conditions** Weather data is collected and processed by scientists. Scientists use sensors to collect weather data. Scientists give weather information to all public sectors which need it. Agriculture is a public sector and it also needs weather data because cultivation depends on the weather condition.
3. **Soil Quality** In smart farming, soil quality is measured by the sensors. For any crop production, soil quality is matters. Sensors can be used to check soil PH level, nutrient value, drainage capacity of the soil, and the direness of soil as shown in FIGURE 5.

HOW IS IOT INTEGRATED WITH AGRICULTURE?

Agriculture integrated with IoT using robots, drones, sensors with analytical tools for monitor the crop field. Physical pieces of equipment are placed on the crop field to capture data. Industrial agriculture is the focus of technology-based smart agriculture (Kumar et al., 2021). Farmers are required by smart farming to employ smart agricultural equipment. The greatest infrastructure for utilizing cutting-edge technologies like big data, cloud computing, and IoT is part of smart farming. Technology is being used in agriculture to track, monitor, automate, and evaluate farm activities. Software and sensors are used to produce smart farming. The growth of intelligent agriculture is made possible by the increase in global population. The demand for crops grows along with each nation's growing population. IoT has made a difference in every industry sector. IoT (Lin et al., 2020) just isn't offering a solution to the issues in agriculture, in fact. Utilizing the most recent technologies, it has altered the paradigm of farming and increased agricultural output. IoT sensors have an unlimited range. Basically, it depends on your demands as to where and how you do it. Nearly every industry uses sensors, including: sensors used in agriculture, businesses, and residences. For instance, sensors assist farmers in agriculture. In agriculture, sensors are used to assess plant performance and fertilizer. Technology is used in tandem with smart farming. Satellites, satellite communications (D. Patil, 2015), sensors, and data analytics are examples of smart agricultural technology. Crop yields are increased via smart farming (D. S. N. Patil & Jadhav, 2019). Several sensors are used in agriculture to collect data. Sensors are utilized for temperature, pH, light, crop damage, water levels, and soil scanning. Among the technologies

Figure 5. Iot Play Different Role to Monitoring Soil

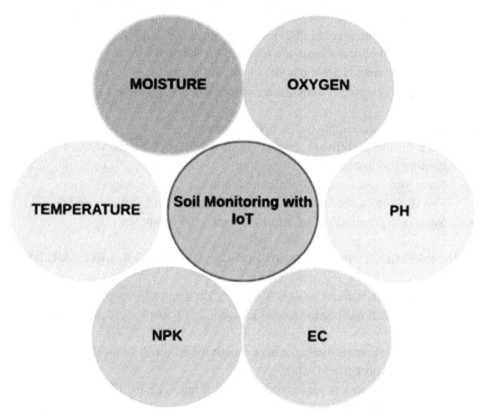

used in telecommunications include networks, GPS, hardware, and software. In smart farming, both hardware and software are crucial. Drones are not the only robots utilized in intelligent farming. Tractors and other agricultural machinery are also equipped with the technologies needed for autonomous driving. Agriculture became independent as a result, requiring less human work. The real procedure is quite simple. AI and location tracking software that makes use of map data are used to program the vehicle's position and speed. In other words, all the GPS and control parts of the machine, including the cameras and sensors, are there.

IOT RESOURCES ARE REQUIRED TO DEVELOPED SMART AGRICULTURE/ SMART FARMING

- Sensors, robots, drones, and cameras are enforced to be put down on a crop field to monitor the crops and get data.

- Smart skill staff is required to implement and control IoT devices.
- Types of equipment that are using in agriculture integrated IoT (Prathibha et al., 2017, Ramesh et al., 2021). It all is expensive and fragile.
- A lot of maintenance costs are required for the hardware.
- Initial investments are too high.
- Computer imaging is done using sensor cameras and drones with the manual operator.
- No need for the supply chain management.
- Each farm has different crop so farm data has to be managed separately because it is not scalable.
- No requirement is need to maintain log info.
- Few difficulties are faced during integration with old implemented devices.
- Some set of operations are required to start the equipment.

INFRASTRUCTURE IS REQUIRED FOR ADOPTING SMART FARMING

- The initial investment is done in sensors, drones, and bots is their setting up.
- The field staff has needed to provide training to operate and management the things.
- Most important requirement is connectivity power. It is used to charge and operate the drones and robots.
- Maintenance cost is required to control things like hardware maintenance, and servicing.
- Internet connectivity is continuously required.

SENSOR

Sensors are physical devices or it is known as a sub-system that is used to come across the events or any changes in the environment and send these change information to the centralized device. A sensor is an electrical device that generates an output signal and is used to detect physical phenomena. A sensor is an electrical equipment, module, device, or subsystem that detects events in its immediate surroundings and sends a signal to a server for further processing. A lot of type sensors are available in market like biometric sensors, radiation sensors, temperature sensors, position sensors, etc. An appropriate task is executed with a particular sensor. Sensors are used to rack up data and then forward to the server or data center or the cloud. For example, Biometric sensor can used to sense biometric, environmental, biological, audible, or visual. Sensor data is further process and analyze to do a specific task.

AGRICULTURE SENSOR

1. **Electrochemical Sensors**: Electrochemical sensors are used to get key information in precision agriculture. It is required to get information regarding PH and soil nutrients. Electrodes sensors detect the specification of the soil.
2. **Location Sensors:** GPS satellites signals are received by the location sensor to determine latitude and longitude. Location sensor is also determines altitude to within feet.
3. **Optical Sensors**: Optical sensors are used to measure soil properties. Optical sensors have two variables data one is soil reflectance and second one is plant color. This data can be aggregated and processed. The moisture content of the soil, clay, and organic matter can also be determined by using optical sensors.
4. **Mechanical Sensors**: Soil compaction and "mechanical resistance" is measured by mechanical sensor. To predict pulling requirements for ground-engaging equipment it is used on large tractors. For example Honeywell FSG15N1A.
5. **Dielectric Soil Moisture**: It is used to access moisture levels from the soil by measuring the dielectric constant.
6. **Airflow Sensors**: It is used to measure soil air permeability. It is capable to make measurements at singular and dynamic location.
7. **Agricultural Weather:** Weather plays a significant role in agriculture. Agriculture weather augur must be referring to all weather elements that are immediately responsible for farm planning or operations. All these elements will differ from place to place and it also varies on the season to season. A weather station is a containment unit that can be shifted at any different locations throughout lengthening fields. A station is made by various devices like sensors, electricity, and required hardware according to crop that is appropriate for the local crops and the climate. Temperature, soil temperature, rainfall, solar radiation, wind speed, humidity, atmospheric pressure, and direction information is processed at predetermined intervals then stored and send this data to a central data logger through a wireless internet connection (Sushanth & Sujatha, 2018).

BENEFITS OF SMART FARMING

Smart agriculture helps reduce the overall cost of crop production and improve the quality and quantity of crop production. Increasing monitoring (Kumar et al., 2021) of crop production should lead to smart agricultural technology. Smart agriculture has reduced crop production waste. In traditional farming, farmers had to do all the work manually, so they didn't have to perform multiple tasks at the same time, but in smart farming, multiple devices are activated at the same time, so multiple tasks. You can perform tasks. Same operations like crop monitoring, mushroom monitoring, water level monitoring, insects monitoring. With the help of smart farming, control of plant diseases in the early stages of crop production is also strengthened, allowing early detection of plant diseases.

A strategy for getting healthy foods to market has been developed using smart farming. Food quality is the main emphasis of intelligent farming. Utilizing characteristics like moisture, fertilizer, or material composition, precision agriculture is employed to control the soil supply. Smart farming has been developed using IoT and cloud technologies (Zamora-Izquierdo et al., 2019), laying the foundation for the "third green revolution." It is a system that is favored for information application and device connectivity. A robotics system, that is. This system combines sensors, actuators, and the Internet of Things.

CONCLUSION

Cultivation will play a huge role in upcoming years for every country. Every country's population has been increased day by day that why every country's government wants to increase crop production to fulfill the basic need of food for its population. To increase agriculture production farmers needs IoT based smart farming. At present, IoT are supporting different domains of farming for different reason such as it is use for time efficiency, soil monitoring, crop monitoring, water management, pest control, moisture monitoring etc. IoT based smart farming reduces investment cost and human efforts, and it enhance remote accessing control over the crop. IoT based smart farming work with latest techniques, which are required minimum efforts to get large production as a result. With the help of latest techniques farmers get information in early stage to save the crops like crop disease, climate condition, water requirement etc. Smart farming work on production quality and quantity, and it is easy as well as cheaper than traditional farming. Smart farming with latest techniques can easily accommodate changing environmental condition, and it use resource in efficiently way. What's more, these technologies are contributing to solutions that extend beyond farms, including pollution, global warming, and conservation.

REFERENCES

A Smart Information System for Public Transportation Using IoT. (2017). *International Journal of Recent Trends in Engineering and Research, 3*(4), 222–230. doi:10.23883/IJRTER.2017.3138.YCHJE

Ahmed, M. S. (2019). Technical Skill Assessment using Machine Learning and Artificial Intelligence Algorithm. *International Journal of Engine Research, 8*(12). Advance online publication. doi:10.17577/IJERTV8IS120109

Ali Hasnain, H. S. A. (2021). Agriculture Monitoring System Using IoT: A Review Paper. *International Journal on Recent and Innovation Trends in Computing and Communication, 9*(1), 1–6. doi:10.17762/ijritcc.v9i1.5452

Anjali, Khangar, & Bhakre. (2018). A review paper on effective agriculture monitoring system using IoT. *International Journal of Modern Trends in Engineering & Research, 5*(3), 15–17. doi:10.21884/IJMTER.2018.5058.YLOGA

Chettri, L., & Bera, R. (2020). A Comprehensive Survey on Internet of Things (IoT) Toward 5G Wireless Systems. *IEEE Internet of Things Journal, 7*(1), 16–32. doi:10.1109/JIOT.2019.2948888

Dinh, T., Kim, Y., & Lee, H. (2017). A Location-Based Interactive Model of Internet of Things and Cloud (IoT-Cloud) for Mobile Cloud Computing Applications. *Sensors (Basel), 17*(3), 489. doi:10.339017030489 PMID:28257067

García, L., Parra, L., Jimenez, J. M., Lloret, J., & Lorenz, P. (2020). IoT-Based Smart Irrigation Systems: An Overview on the Recent Trends on Sensors and IoT Systems for Irrigation in Precision Agriculture. *Sensors (Basel), 20*(4), 1042. doi:10.339020041042 PMID:32075172

Harendra Negi, S. C. (2020). Smart Farming using IoT. *International Journal of Engineering and Advanced Technology, 8*(4S), 45–51. doi:10.35940/ijeat.D1015.0484S19

Islam, N., Rashid, M. M., Pasandideh, F., Ray, B., Moore, S., & Kadel, R. (2021). A Review of Applications and Communication Technologies for Internet of Things (IoT) and Unmanned Aerial Vehicle (UAV) Based Sustainable Smart Farming. *Sustainability, 13*(4), 1821. doi:10.3390u13041821

Jayaraman, P., Yavari, A., Georgakopoulos, D., Morshed, A., & Zaslavsky, A. (2016). Internet of Things Platform for Smart Farming: Experiences and Lessons Learnt. *Sensors (Basel), 16*(11), 1884. doi:10.339016111884 PMID:27834862

Jha, K., Doshi, A., Patel, P., & Shah, M. (2019). A comprehensive review on automation in agriculture using artificial intelligence. *Artificial Intelligence in Agriculture, 2*, 1–12. doi:10.1016/j.aiia.2019.05.004

K, A., & S, N. S. R. (2019). Analysis of Machine Learning Algorithm in IOT Security Issues and Challenges. *Journal of Advanced Research in Dynamical and Control Systems, 11*(9), 1030–1034. doi:10.5373/JARDCS/V11/20192668

Kasar, M. V. V. (2019). Smart Bins Concept Implementation in India- Garbage Monitoring System using IOT. *International Journal for Research in Applied Science and Engineering Technology, 7*(6), 1939–1942. doi:10.22214/ijraset.2019.6326

Khoa, T. A., Man, M. M., Nguyen, T. Y., Nguyen, V., & Nam, N. H. (2019). Smart Agriculture Using IoT Multi-Sensors: A Novel Watering Management System. *Journal of Sensor and Actuator Networks, 8*(3), 45. doi:10.3390/jsan8030045

Kumar, A., Kumar, A., Singh, A. K., & Choudhary, A. K. (2021). IoT Based Energy Efficient Agriculture Field Monitoring and Smart Irrigation System using NodeMCU. *Journal of Mobile Multimedia.* doi:10.13052/jmm1550-4646.171318

Lin, J., Long, W., Zhang, A., & Chai, Y. (2020). Blockchain and IoT-based architecture design for intellectual property protection. *International Journal of Crowd Science, 4*(3), 283–293. doi:10.1108/IJCS-03-2020-0007

Machine Learning Prediction Analysis using IoT for Smart Farming. (2020). *International Journal of Emerging Trends in Engineering Research, 8*(9), 6482–6487. doi:10.30534/ijeter/2020/250892020

Malik, I., & Tarar, S. (2021). Cloud-Based Smart City Using Internet of Things. *Integration and Implementation of the Internet of Things Through Cloud Computing,* 133–154. doi:10.4018/978-1-7998-6981-8.ch007

Mekala, M. S., & Viswanathan, P. (2017). A Survey: Smart agriculture IoT with cloud computing. *2017 International Conference on Microelectronic Devices, Circuits and Systems (ICMDCS).* 10.1109/ICMDCS.2017.8211551

Patil, D. (2015). Tumor Size Processing using Smart Phone. *International Journal on Recent and Innovation Trends in Computing and Communication, 3*(2), 785–788. doi:10.17762/ijritcc2321-8169.150275

Patil, D. S. N., & Jadhav, M. B. (2019). Smart Agriculture Monitoring System Using IOT. *IJARCCE, 8*(4), 116–120. doi:10.17148/IJARCCE.2019.8419

Prathibha, S. R., Hongal, A., & Jyothi, M. P. (2017). IOT Based Monitoring System in Smart Agriculture. *2017 International Conference on Recent Advances in Electronics and Communication Technology (ICRAECT).* 10.1109/ICRAECT.2017.52

Ramesh, M., Vijay Kumar, G., Suresh Babu, B., Boopathi, R., Sreekanth, C., Muthukumar, P., & Padma Suresh, L. (2021). Exploration or Multipurpose Electric Vehicle for Agriculture Using IOT. *Tobacco Regulatory Science*, *7*(5), 3844–3852. doi:10.18001/TRS.7.5.1.157

Rathod, N. (2020). Smart Farming: IOT Based Smart Sensor Agriculture Stick for Live Temperature and Humidity Monitoring. *International Journal of Engine Research*, *V9*(07). Advance online publication. doi:10.17577/IJERTV9IS070175

Ryu, S., Kim, K., Kim, J. Y., Cho, I. K., Kim, H., Ahn, J., Choi, J., & Ahn, S. (2022). Design and Analysis of a Magnetic Field Communication System Using a Giant Magneto-Impedance Sensor. *IEEE Access: Practical Innovations, Open Solutions*, *10*, 56961–56973. doi:10.1109/ACCESS.2022.3171581

S., K. (2018). IoT in Agriculture: Smart Farming. *International Journal of Scientific Research in Computer Science, Engineering and Information Technology*, 181–184. doi:10.32628/CSEIT183856

S., L., & B., H. (2018). Design and Implementation of IOT based Smart Security and Monitoring for Connected Smart Farming. *International Journal of Computer Applications, 179*(11), 1–4. doi:10.5120/ijca2018914779

Sushanth, G., & Sujatha, S. (2018). IOT Based Smart Agriculture System. *2018 International Conference on Wireless Communications, Signal Processing and Networking (WiSPNET)*. 10.1109/WiSPNET.2018.8538702

Tapakire. (2019). IoT based Smart Agriculture using Thingspeak. *International Journal of Engineering Research And, 8*(12). doi:10.17577/IJERTV8IS120185

Zamora-Izquierdo, M. A., Santa, J., Martínez, J. A., Martínez, V., & Skarmeta, A. F. (2019). Smart farming IoT platform based on edge and cloud computing. *Biosystems Engineering*, *177*, 4–17. doi:10.1016/j.biosystemseng.2018.10.014

Chapter 7
AUTOHAUS:
An Optimized Framework for Secure and Efficient Parking

Shanu Sharma
ⓘ https://orcid.org/0000-0003-0384-7832
Department of Computer Science and Engineering, ABES Engineering College, India

Misha Kakkar
ⓘ https://orcid.org/0000-0002-2061-0477
Department of Computer Science and Engineering, ASET, Amity University, Noida, India

Tushar Chand Kapoor
Department of Computer Science and Engineering, ASET, Amity University, Noida, India

Rishi Kumar
Universiti Teknologi Petronas, Malaysia

ABSTRACT

In this chapter, an optimized parking framework, AUTOHAUS, is presented that focuses on three aspects (i.e., automation, security, and efficient management of parking space). A combination of advanced technologies is used to design the proposed framework. AUTOHAUS provides two ways of security implementation such as authorized QR codes and OTPs (one-time passwords). Furthermore, for efficient management of parking spaces, the still images of the front and side view of the car are used to extract the license plate and size of the car for effective allotment of parking space based on the size of the car. This proposed system can reduce human effort to a great extent and can also be used as a path-breaking technique in parking and storage management.

DOI: 10.4018/978-1-6684-4991-2.ch007

INTRODUCTION

Nowadays there is a tremendous rise in the automobile industry and the need for an efficient and secure parking system is of great concern today. This great increase in this industry has led to an abundance of cars which leads to a shortage of parking spaces (Jusat et al., 2021). There are various automated parking systems available, but none of those systems has the ability to manage space efficiently (Mathijssen & Pretorius, 2007); (Serpen & Debnath, 2019). Hence, there is a need for a mechanism that will replace the current system which doesn't have the ability to manage special resources. The current automated parking systems have shown a great deal in the evolution of automation technology but technology can always be optimized and improved (Pala & Inanc, 2007). Seeing these current systems and the issue of the decreasing parking spaces, a solution comes to light that is based on the type and the length of the car these spaces and can be efficiently managed and utilized. Also seeing today's users are very convenient using mobile devices for everyday tasks. The integration of the parking system with mobile device brings additional functionality to the system which not only provides ease of access to the users by finding the parking system and automatic payments, but it can also provide a layer of security that only the mobile device will be able to control the entry and exit of the car from the system.

Motivated by the mentioned problem domain and through the development of new technologies, here an efficient and automatic parking model is presented, where two main parameters are presented i.e., security of cars and the efficient use of parking space. When the car enters the parking system based on its type and size a parking spot is assigned to the car, this is based on the theory that three small cars take up almost the same space as two sedan cars. Furthermore, with this, some security features are also implemented and an app was developed to speed up the parking process.

This proposed framework is a solution to various parking problems that a man faces daily, first it caters to the biggest problem of efficient parking spaces which is a major concern in today's world, especially where there is a scarcity of space. Secondly, the system's interface has two options for making an entry which includes scanning of the QR code option for the user who has a mobile app pre-installed, in this option the user has to just scan the QR, and the rest is done automatically at the system's end, and at the time of existing user just has to show the generated code at mobile in the system for the successful exit. The second option is for the user who doesn't have the mobile application installed on their mobile device, this is option is an OTP password system an OTP (One Time Password) is sent to the user's mobile number and the user has to enter the password at the time of entry, and at the time of exit, the user has to put in another password sent to the same mobile device to exit the system. The mobile application of this system will have some additional features which will cater to some problems like finding the parking system and paying for the services. The application will have an option to automatically pay for the parking services and it will have another tab that will have the option to find the parking system for the users who are unaware of the location of the parking or are new to the area, the application will help them find and navigate to the nearest parking system.

BACKGROUND

Vehicle parking is one of the major issues, Indian cities are currently facing. Due to the rapidly expanding population, more cars are added continuously in the restricted public areas, and thus the majority of Indian cities are facing traffic congestion and parking-related issues. Developing an advanced and optimized parking system is always a key issue among researchers and developers. With the advancement and incorporation of new technologies, developers are trying to develop more advanced methods to solve parking issues more effectively.

Various novel approaches have been successfully implemented. One can see a multilevel parking system in metro cities, where the optimized utilization of vertical parking space can be seen (Mathijssen & Pretorius, 2007). These types of multilevel parking system have made the parking situation a little better, but still, a lot of issues exists such as waiting time, spacing issues, searching for parking space, etc. Various smart parking-related systems were also developed to solve the existing parking-related issues. Technologies of all kinds are being employed to solve the parking issues in public areas. For instance, RFID technology can automate the payment system and shorten the time required for vehicle check-in and check-out. Similar information can be gathered via wireless sensors, such as parking duration, available slot, payment details, directional details, etc., which will benefit drivers and ease parking woes (Pala & Inanc, 2007). Given the growing number of parking issues in both established and emerging nations, smart parking is a niche market in which many businesses are currently making significant investments. IoT and advanced technologies such as AI, ML can provide more efficient and optimized solution to the current issue (Jog et al., 2015); (Hongyan, 2011). Some of the related systems are discussed here to show the need of the more optimized solution for the mentioned issue.

A Car Parking Framework (CPF) based on IoT technology is suggested by Karbab et al. (2015). The framework integrates a radio frequency identification system with a networked sensor and actuator system to manage parking spaces automatically (RFID). The CPF offers security, parking lot retrieval, payment options, and vehicle guiding. Instead of standard nodal communication, the system employs a hybrid communication technique. As a result, the system uses little energy and is inexpensive to implement. A protected parking reservation system using the Global System for Mobile (GSM) technology is proposed by (Rahayu & Mustapa, 2013). A parking space monitoring module and a security reservation module make up the system. The reservation of a specific parking lot is handled by the security reservation module. To reserve a parking space, the user must send a Short Message Service (SMS) containing detailed instructions. The parking lot monitoring module shows the user animation of the layout showing the occupancy status of the parking spaces, allowing them to select a parking lot for reservation. A password is generated by the system and is necessary at entry and exit points. Further the use of computer vision technology for solving parking related issue is explored in (Masmoudi et al., 2014), where the system offers an outdoor parking service that can locate vacant parking spaces in real-time and transmit the address to the driver for vehicle navigation. To deal with the security issues in parking system, an IoT-based system is proposed by (Singh et al., 2019), which uses a two-way security using face recognition and license plate recognition-based approach to offer a key-based parking reservation system that guarantees the right person will be assigned the parking space

Although a lot of systems have been proposed to date (Serrão & Garrido, 2019); (Fahim et al., 2021), still their real-time using is missing due to a range of implementation and cost issues. Motivated by the issues discussed, in this paper, an automated parking framework is proposed which focuses on three main factors i.e., complete automation, security, and efficient utilization of space. In the proposed approach, complete automation is proposed where users can enter the system through QR code or OTP-based authentication method, in parallel using the front and side view of the car, the license plate number and size of the car will be calculated using a computer vision technology. Further, an optimized slot according to the size of the car will be allocated to the unique license plate number. The proposed approach can be used in both fully automatic and semi-automatic way, where users can park the car themselves or it can be parked by the automated parking machine at the calculated optimized slot. The user can simply check out through the system by showing the QR code at the exit. After successful authentication, the car will be delivered to the user at the exit point. The proposed system deals with the security and space-related issues and provide a complete hassle-free solution to the parking issue.

Figure 1. Flow diagram of the proposed parking framework

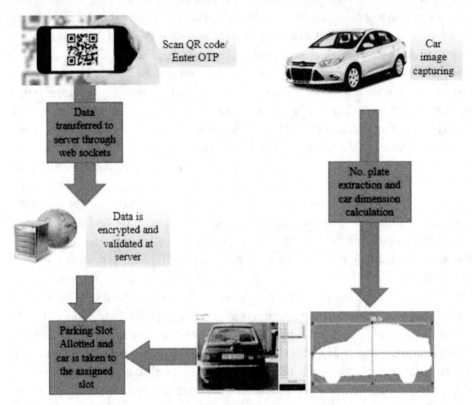

AUTOHAUS: PROPOSED PARKING FRAMEWORK

System Designing

The proposed AUTOHAUS parking framework is a model for providing an effective parking framework to the users, which utilizes the parking space effectively and an end-to-end automated and secure parking system using a combination of advanced technologies. The workflow of the proposed parking framework is presented in Figure 1. Various steps of working of the system is discussed below.

- The system consists of two types of authentications for secured entry of cars into the parking infrastructure. The first one is QR code-based authentication, and the other is One-time password (OTP) based authentication. Both methods can be accessed by the user through the mobile application. After the successful authentication, the data of the user will be transferred to the web server using web sockets which create an open-bidirectional communication channel. After the data is sent to the server, the server decrypts the data received for validation and after the validation, two parallel processes are started for finding the correct slot for parking

- The two parallel processes are then executed to extract the information from the images of the car. The still images of the car are extracted using installed cameras in two directions, one contains the front view of the car to extract the license plate no. of the car and the other one contains the side view of the car to estimate the size of the car.
- The extracted data from the above-mentioned steps are then used by the system to assign the specific slot of parking to the particular car based on its size.

Hardware/Software Requirements

Software requirements

- Operating System: Windows 10/8.1/8/7
- Processors: Any Intel 64-x86 processor
- RAM: 2 GB is recommended
- Disk Space: 4–6 GB for a typical installation

Hardware requirements

- Arduino Board
- Two Cameras
- Servo Motors

System Implementation

QR Code Authentication

As mentioned in the system designing section, the proposed system consists of two types of authentications i.e., QR code-based authentication and OTP-based authentication. The complete process of QR code-based authentication in the proposed parking framework is discussed below.

An encrypted QR code will be installed in the proposed parking system, and the user will be able to scan the QR code using any scanning-enabled mobile application. The QR Code is designed in such a way that the data it carries is encrypted by 128-bit encryption using AES algorithm. Code encryption is used to ensure that there

Figure 2. Encryption process in QR code-based authentication (Al-Ghaili et al., 2020)

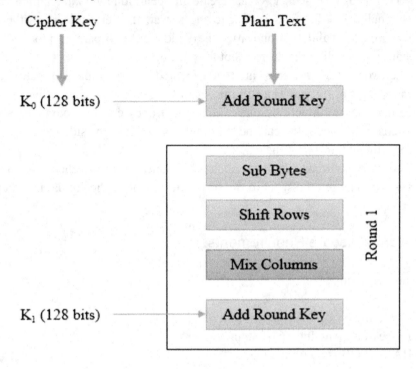

is no security breach when some tries to read this code using other methods which makes it a more reliable and secure method of authentication for entering the system.

Encryption Process (Naser et al., 2020): The encryption process contains four sub processes; it starts with Sub bytes which is the byte substitution method that gives out the result into 4x4 matrix form. Then comes the shift rows in which all the rows are shifted to the left which further includes some sub steps. Then in Mix Columns the transformation of the matrix is done using distinctive mathematical functions, after that during the Add round Key step, the 16 bytes matrix is now considered to be 128 bits which is the last round, after that the output comes as the cipher text, the Flow of Encryption process of the AES algorithm is presented in Figure 2:

Decryption Process: The decryption process is the reverse of the encryption process where each step of encryption process is implemented in the reverse direction to generate the original text (Al-Ghaili et al., 2020).

QR Code Generation (Naser et al., 2020): The data that needs to be represented in the QR Code is encrypted at the server side and sent to the client, then using that data JavaScript prints QR code on the client side with the help of an open source library qrcode.js. The sample code for QR Code generation is presented in Figure 3.

Figure 3. Sample QR Code Generation Code

```
<div id="qrcode"></div>

<script type="text/javascript">
var qrcode = new QRCode("test", {
    text: "Sample QR Code",
    width: 128,
    height: 128,
    colorDark : "#000000",
    colorLight : "#ffffff",
    correctLevel : QRCode.CorrectLevel.H
});
</script>
```

License Plate Recognition

The recognition of number plates is a complex process, for that a special open-source tool is used known as JavaAnpr, which is Java Automatic Number Plate Recognition System (Lubna et al., 2021). The steps that are followed by this tool are described as follows:

a) Rank Filtering and Edge Detection: Figure 4 shows, how the original image is first vertically and horizontally rank filtered then horizontal and vertical edge detection is done using sobel edge detection method (Lubna et al., 2021).

b) Two Phase image analysis: In the first phase, the detection of the portion of the number plate which covers the wider area is covered (Lubna et al., 2021). Then when it comes to the second phase the result from the first is skewed and then processed as shown in Figure 5.

c) Segmentation of the Number plate: Number plate is segmented using the horizontal projection, adaptive thresholding is done in order for the separation of the dark foreground from the comparatively light background as shown in Figure 6.

Figure 4. Rank filtering and edge detection

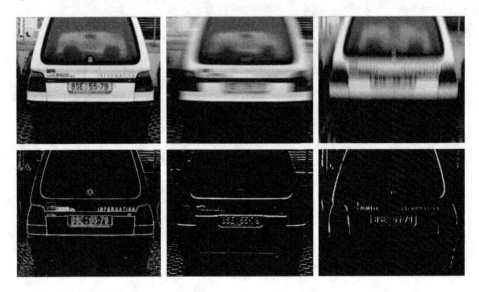

d) Feature Extraction: To extract the specific characters in a number plate, a pixel matrix is created as show in Figure 7, then the number of loops are determined as shown in Figure 8.

Figure 5. Two phase plate clipping

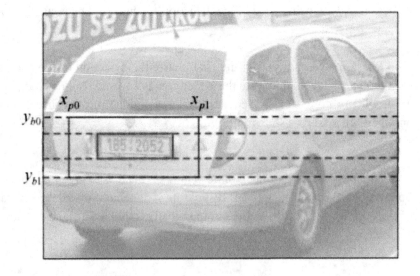

e) Recognition of characters: After feature extraction step, the specific characters are then recognized which results into the recognized number plate.

Figure 6. Result of application of adaptive Thresholding on number plate

Figure 7. The "pixel matrix" of specific character

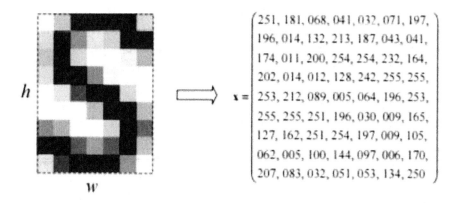

Figure 8. Calculation of number of loops

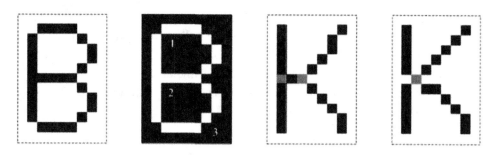

Figure 9. Image Conversion from RGB color space to L*a*b* color space (a) Input Image (b) Converted Image

Car Size Estimation

To estimate the size of the car from its image, first, the car part is segmented from the background, then the length of the car is calculated. Various steps in achieving this are discussed below:

Segmentation of Car from Background: For the segmentation of the car from the image background, the color-based segmentation is used which is done by using K-Means Clustering algorithm (Na et al., 2010). K-Means clustering is a powerful algorithm and can be easily implemented using requires "Statistical and Machine Learning Tool Box of MATLAB". The various algorithmic steps of implementing K means Clustering algorithm on the colored images is described as follows:

a)   Image Conversion into L*a*b* color space from RGB color Space: Converting the image into L*a*b* color space from RGB colors: In this step, the image is converted into "L*a*b*" color space, which comprises of "luminosity layer L*", "a*" layer which is the chromatic layer, and is used for indicating the falling point of the color on the "red-green" axis (Li & Wu, 2012). Another layer "b*" which is the chromatic layer, is used for indicating the falling point of the color on the "blue-yellow" axis. The output of the color conversion is presented in Figure 9.

b)   K-Means Clustering: The K-means clustering is then applied to separate object from the background. K means clustering follows the principle that every object is having a unique location in 3-d space. The main task is to discover partitions in such a way that objects of the same clusters are kept as close as possible and the objects of the other clusters are as farthest as possible (Na et al., 2010).

Figure 10. Image labelled by clustering index

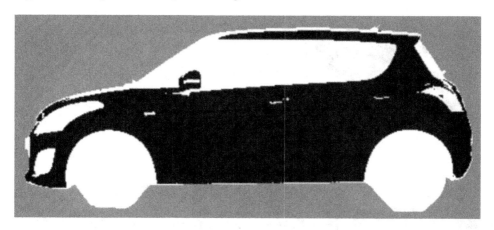

c) Labeling the pixels in the image after K means clustering: After the K Means Clustering is applied every pixel in the image is labeled with it is corresponding clustering index, as shown in Figure 10.

d) Car Segmentation: With the help of K-Means clustering, different clusters were extracted, to extract the cluster contains car as shown in Figure 11.

Car Size Calculation: Measuring the length of the car is done by using an open-source tool OpenCV which is fully packed with computer vision algorithms. Here OpenCV on the python programming language to calculate the length of the car. There are various steps involved in the calculation of the length of the car which is described as follows:

• First, various required packages were imported into the python program, the most important package is "imutils". The input image is loaded and the important preprocessing steps were implemented. First, the image is converted into grayscale, followed by smoothening of the image using Gaussian filter as shown in Figure 12

Figure 11. Different clusters of the input image

Figure 12. Segmented Image conversion into Grayscale

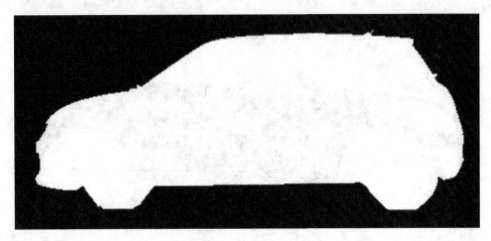

and Figure 13 respectively. Then edge detection is performed using a combination dilation and erosion as presented in Figure 14 (Canny, 1986).

- Then a midpoint () function is implemented, which is used for computing the midpoints for the major and minor axis. In this step a loop is started in order to decide that if a contour is not sufficiently large it should not be taken into consideration assuming that as noise (Manchanda & Sharma, 2017). If the contour is found to be considerably large then the rotated bounding box is computed, with surrounding midpoints as shown Figure 15.

Figure 13. Smoothening of Image using Gaussian Filter

Figure 14. Output of Dilation Erosion on filtered Image

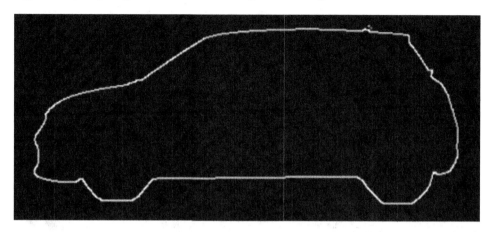

- Finally, the length of the car is calculated using the Euclidean distance between points on the extracted bounding box as mentioned in figure 16.

Parking Slot Assignment

After successful authentication from the QR code, and through the extracted data from the car image i.e., car length and license plate, the assignment of slot is made. The availability of the slots is derived from a custom-made database as shown in Figure 17. The slot assigning algorithm is designed in such a way that it caters to

Figure 15. Bounding box Estimation

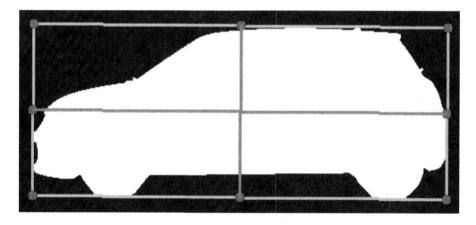

Figure 16. Car Length Calculation

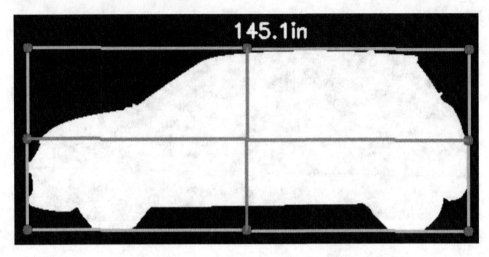

every possibility to assign a slot so that no slot is wasted and the maximum number of cars can be parked in the system.

AUTOHAUS: SYSTEM INTERFACE AND USE

In this section, various results obtained during the testing of the model are presented and discussed. The front-end startup interface of the AUTOHAUS parking system is shown in Figure 18, in which there are two options for making an entry into the parking system, first being the QR Code method and second being the OTP method.

Figure 17. Database Snap for Slot Allocation

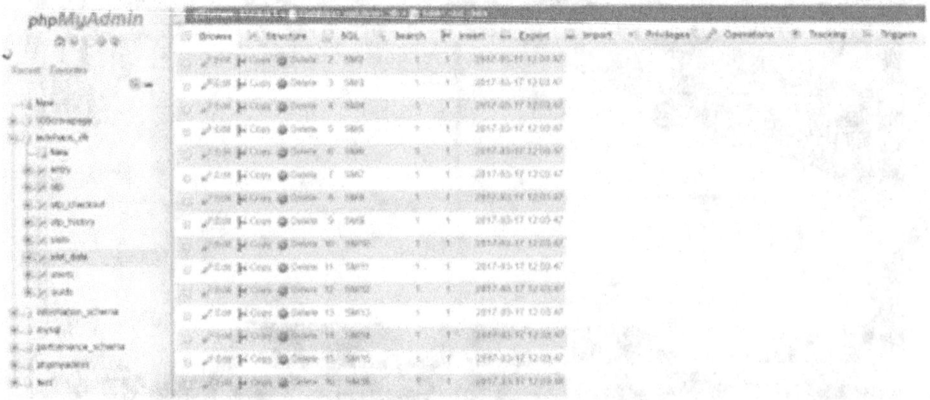

Figure 18. Front End Startup Interface of the AUTOHAUS parking system

After the successful selection of the provided options by the user, the second phase of the interface starts as per the selection. If the user selects the QR Code method, an encrypted QR code is generated as shown in Figure 19, in which a unique code is displayed for each entry. The QR code can be scanned through a mobile application. The front end of the mobile application is presented in Figure 20, which also performs the task of decrypting the code displayed on the system. After successful authentication of the user, a parking slot will be assigned to the user according to the length of the car as shown in Figure 21.

Further, if the user selects an OTP method for entering into the parking system, an option to enter a mobile number is displayed as shown in Figure 22. After successful entering of mobile number, and an OTP is sent to the mobile device as shown in Figure 23.

After successful authentication of the user into the car parking system through any of the provided options i.e., OTP based authentication or QR code-based authentication, the second phase of the system starts which is the assignment of parking slot according to the length of the car. This system runs in parallel to authentication phase when the car enters the system. First, an automatic license plate recognition is performed on the front view of the car as shown in Figure 24. Second, the length of the car is calculated through the side view of the car. The side view of the car is presented in Figure 25, which is then used to calculate the length of the car as per its size using the approach mentioned in the previous section. Based on this calculation a parking slot is assigned to the user on the name of its license number.

Figure 19. Encrypted QR Code

After successful parking of the car at the assigned parking slot, a user can check out from the system using the same two options i.e., OTP or QR code-based checkout. The checkout interface of the parking system is shown in Figure 26.

The user will have to show the QR code at the exit of the system, which will provide the car and the parking slot information to the backend. This QR code will be used to authenticate the user, and after successful authentication, the car will be delivered to the user.

CONCLUSION

In this chapter, an automated parking framework is proposed to tackle the space, security, and automation-related issues during the parking of the car. The proposed AUTOHAUS parking system provides a hassle-free automated facility to the user. It is observed that the space provided for a big car can be used to park two small cars. Based on this space-related concept, here the size of the car is used to assign a specific parking slot to the car. Furthermore, to provide a completely automated parking environment to the user, here two methods have been proposed to authenticate the user i.e., OTP-based authentication and QR code-based authentication. The user can check in and check out from the proposed automated parking system using any of these methods. The system provides a space-efficient parking slot to the car, further at the time of checkout the same authentication method can allow the hassle-free successful delivery of the car to the user. In this paper, the prototype of the proposed parking system is presented, which shows its real-time efficacy and

Figure 20. Front End of the Mobile Application

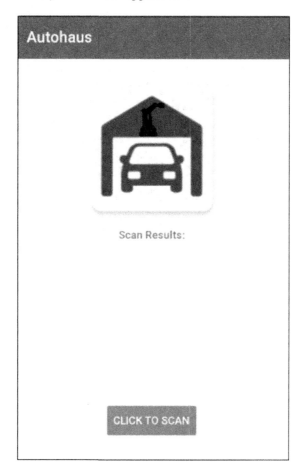

applicability. This type of automated parking system is much-needed system in the world of advanced technology.

Figure 21. Confirmation after Successful Authentication

Figure 22. OTP based authentication method

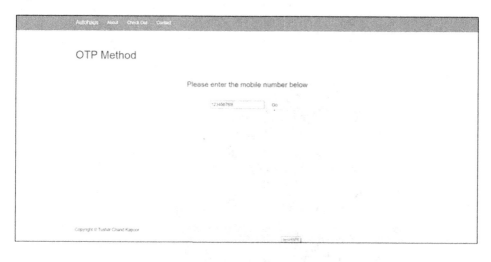

Figure 23. Successful authentication through OTP

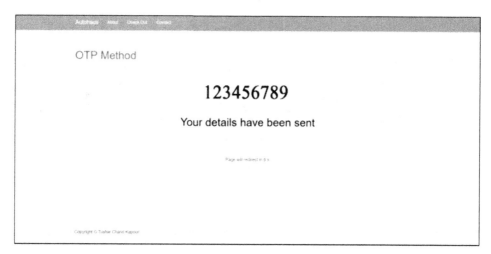

Figure 24. License plate recognition through front view of the car

Figure 25. Side view of the Car

Figure 26. Front End of the Checkout Interface

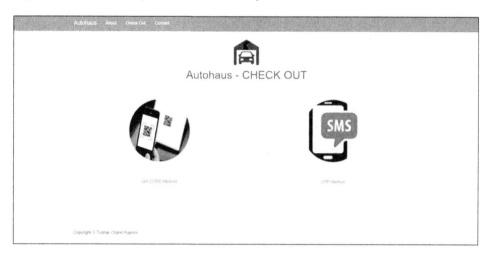

REFERENCES

Al-Ghaili, A. M., Kasim, H., Othman, M., & Hashim, W. (2020). QR code based authentication method for IoT applications using three security layers. *TELKOMNIKA*, *18*(4), 2004. doi:10.12928/telkomnika.v18i4.14748

Canny, J. (1986). A Computational Approach to Edge Detection. *IEEE Transactions on Pattern Analysis and Machine Intelligence*, *PAMI-8*(6), 679–698. doi:10.1109/TPAMI.1986.4767851 PMID:21869365

Fahim, A., Hasan, M., & Chowdhury, M. A. (2021). Smart parking systems: Comprehensive review based on various aspects. *Heliyon*, *7*(5), e07050. doi:10.1016/j.heliyon.2021.e07050 PMID:34041396

Hongyan, K. (2011). Design and Realization of Internet of Things Based on Embedded System Used in Intelligent Campus. *International Journal of Advancements in Computing Technology*, *3*(11), 291–298. doi:10.4156/ijact.vol3.issue11.37

Jog, Y., Sajeev, A., Vidwans, S., & Mallick, C. (2015). Understanding Smart and Automated Parking Technology. *International Journal of U- and e-Service Science and Technology*, *8*(2), 251–262. doi:10.14257/ijunesst.2015.8.2.25

Jusat, N., Zainuddin, A. A., Sahak, R., Andrew, A. B., Subramaniam, K., & Rahman, N. A. (2021, August). Critical Review In Smart Car Parking Management Systems. *2021 IEEE 7th International Conference on Smart Instrumentation, Measurement and Applications (ICSIMA)*. 10.1109/ICSIMA50015.2021.9526322

Karbab, E., Djenouri, D., Boulkaboul, S., & Bagula, A. (2015, May). Car park management with networked wireless sensors and active RFID. *2015 IEEE International Conference on Electro/Information Technology (EIT)*. 10.1109/EIT.2015.7293372

Li, Y., & Wu, H. (2012). A Clustering Method Based on K-Means Algorithm. *Physics Procedia*, *25*, 1104–1109. doi:10.1016/j.phpro.2012.03.206

Lubna, M., Mufti, N., & Shah, S. A. A. (2021). Automatic Number Plate Recognition:A Detailed Survey of Relevant Algorithms. *Sensors (Basel)*, *21*(9), 3028. doi:10.339021093028 PMID:33925845

Manchanda, S., & Sharma, S. (2017). Extraction and Enhancement of Moving Objects in a Video. *Advances in Computer and Computational Sciences*, 763–771. doi:10.1007/978-981-10-3770-2_72

Masmoudi, I., Wali, A., Jamoussi, A., & Alimi, A. M. (2014). Vision based System for Vacant Parking Lot Detection: VPLD. *Proceedings of the 9th International Conference on Computer Vision Theory and Applications*, 526–533. 10.5220/0004730605260533

Mathijssen, A., & Pretorius, A. J. (2007). Verified Design of an Automated Parking Garage. *Formal Methods: Applications and Technology*, 165–180. doi:10.1007/978-3-540-70952-7_11

Na, S., Xumin, L., & Yong, G. (2010, April). Research on k-means Clustering Algorithm: An Improved k-means Clustering Algorithm. *2010 Third International Symposium on Intelligent Information Technology and Security Informatics*. 10.1109/IITSI.2010.74

Naser, M. A. U., Jasim, E. T., & Al-Mashhadi, H. M. (2020). QR code based two-factor authentication to verify paper-based documents. *TELKOMNIKA*, *18*(4), 1834. doi:10.12928/telkomnika.v18i4.14339

Pala, Z., & Inanc, N. (2007, September). Smart Parking Applications Using RFID Technology. *2007 1st Annual RFID Eurasia*. doi:10.1109/RFIDEURASIA.2007.4368108

Rahayu, Y., & Mustapa, F. N. (2013). A Secure Parking Reservation System Using GSM Technology. *International Journal of Computer and Communication Engineering*, 518–520. Advance online publication. doi:10.7763/IJCCE.2013.V2.239

Serpen, G., & Debnath, J. (2019). Design and performance evaluation of a parking management system for automated, multi-story and robotic parking structure. *International Journal of Intelligent Computing and Cybernetics*, *12*(4), 444–465. doi:10.1108/IJICC-02-2019-0017

Serrão, C., & Garrido, N. (2019). A Low-Cost Smart Parking Solution for Smart Cities Based on Open Software and Hardware. *Lecture Notes of the Institute for Computer Sciences, Social Informatics and Telecommunications Engineering*, 15–25. doi:10.1007/978-3-030-14757-0_2

Singh, A. K., Tamta, P., & Singh, G. (2019). Smart Parking System using IoT. *International Journal of Engineering and Advanced Technology*, *9*(1), 6091–6095. doi:10.35940/ijeat.A1963.109119

Chapter 8
IoT–Enabled Smart Homes:
Architecture, Challenges, and Issues

Indu Malik
ABES Engineering College, India

Arpit Bhardwaj
BML Munjal University, India

Harshit Bhardwaj
Galgotias University, India

Aditi Sakalle
Gautam Buddha University, India

ABSTRACT

The smart home achieved commerciality in the previous decades because it has increased comfort and quality of human life. Using AI and IoT technology, humans get notifications about the unplaced (not having or assigned to a specific place) things or devices at home. In this chapter, the authors introduce the concept of a smart home with the integration of IoT (internet of things), cloud, and AI. A smart home can be controlled by smartphones, tablets, or PC. The smart home is a collection of smart devices; smart devices have sensors and actuators. All these devices are connected, and they are controlled by the central unit. The sensor is used to collect data. The smart home is a net of sensors, different types of sensors used to create a smart home. Particular sensors perform particular tasks like biometric sensors used for security. In this chapter, the authors discuss remote access of devices, increased computation power, and data storage. Home automation gives you access to an on and off home light, lock and unlock the door, and switch on and switch off the AC and TV.

DOI: 10.4018/978-1-6684-4991-2.ch008

INTRODUCTION

The smart home (A et al., 2017) is a collection of merged technologies such as IoT, cloud computing, and AI. Instead of this, these technologies are responsible to manage local and central computing with the optimized resource. Masterly smart home, IoT, cloud, and AI are basic components for our proposed smart home. Each component is a core attribute to create a smart home. IoT is used for internet connection, sensors, and remote access of smart devices through central control unit (PC, mobile, tablet). In Day to Day, life humans need smart devices, which make human life easier. Sensors could be attached to any electronic device such as the doorbell, air conditioner, light, etc. The smart home is also a part of a smart computer or intelligent computer which is why it embeds with home devices to control and monitor home appliances' functionality. Cloud computing is used for storage space, scalable power computing, and applications for developing, and maintaining home services, with the help of cloud computing home devices can be accessed anywhere. The smart home is used to access control over the devices, to get security, energy efficiency, and convenience. Smart devices provide convenience and saving of people time, money, and energy. The smart home is controlled by a smartphone or tablet. In this chapter, we introduce what is Smart home and why a Smart home is needed? How we can create a smart home using IoT, AI, and the cloud. Nowadays AI and IoT (Alaa et al., 2017) technology is used to create a smart home.

Home Automation

Before you delve deep into smart home technologies, you should be known about home automation. Home automation is an ecosystem. It refers to the automatic control of the electronic devices, home activity, and appliances of the home. In simple terms, home automation is a system that gives the features to control utilities and functionality of the home remotely through the internet.

How Does Home Automation Work?

Home automation is a collection of hardware, software, and electronic interface, which works to integrate all devices with another one through the internet. Each electronic device has sensors and these are connected through the internet. The House owner could manage or control all devices through a smartphone or tablet house owner's location does not matter, it is at home or a milestone away from the home. Home automation gives you access to an ON and OFF home light, lock and unlock the door, and switch on and switch off the AC and TV.

Figure 1. Home Automation Components

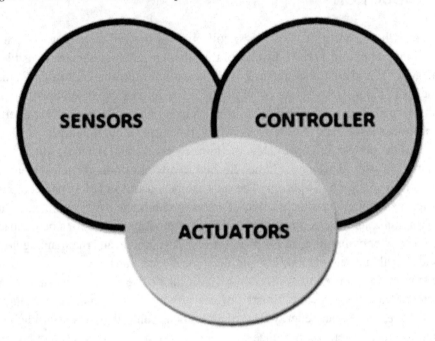

Home Automation has three basic components as shown in Figure 1. These components are

- Sensors
- Controllers
- Actuators

Sensors: Sensors are used to sense the activity; it used to sense or detect smoke, motion, and temperature. Home automation adjusts home settings according to user preferences.

Controllers: Basically, controller is a device that operates all the functionality of smart devices such as smartphones, tablets, and computers.

Actuators: Actuators may be light switches or motors. Actuators control the actual mechanism or functions of the home automation.

SMART HOME

"Smart homes" are those homes that have electronic devices with a sensor as well as a security system and all these devices are connected through the internet and it operates by the owner using smart mobile, tablet, and computer". Smart home technology, commonly referred to as home automation or domotics (from the Latin word "domus" in house), gives homeowners protection by allowing them to control their smart devices using a smartphone app. provides comfort, convenience, energy-saving features, or other network hardware. As part of the Internet of Things (IoT), smart home systems and devices commonly work together to communicate consumer usage data and automate tasks based on homeowner preferences. The main access point for all connected devices in smart homes might be a smartphone, tablet, laptop, or game console. Additionally, electrical devices such as door locks, TVs, thermostats, home monitors, cameras, lights, and even refrigerators may be controlled by home automation systems. The technology is implemented on mobile or other connected devices and allows users to specify timeframes for when certain changes should take effect.

The smart home is a term that denotes a technical system for a normal home. This technical system has a collection of IoT, sensors, AI, software, and automated process, Using the technical system user is capable to control or monitor home functionality remotely instead of physically. Home security has been enhanced through the smart home. The objective of smart home is to make human life easy. If a person has gone to the office and is doing office work, suddenly, he recalls he has forgotten to switch off his home AC. Now it can switch off its home AC through a smart application that is already installed on its phone. It is an example of a smart home. Smart home architecture is shown Figure 2. All smart devices can be addressed within their radius through the control unit.

- For smart home, the first one is an internet connection is required.
- To control and monitor smart home devices, a PC, tablet, or smartphone is required.
- For every smart device which is used at home, its corresponding application software must be installed in a control unit such as a PC, smartphone, or tablet. Smart devices are controlled through the program.
- To remote access, a smart device's wi-fi connection is required.

Smart Home Appliances

The smart home is a collection of appliances such as an air conditioner, washing machine, dishwasher, gas, light, oven, and coffee maker. All these devices are made using sensors, IoT, Internet, and AI. All these devices are controlled through Central Control Unit. CCU is the heart of the smart home. All smart home components are connected and it's controlled through maybe a Personal Computer (PC), smartphone,

Figure 2. Smart Home Architecture with Internet and Cloud Data

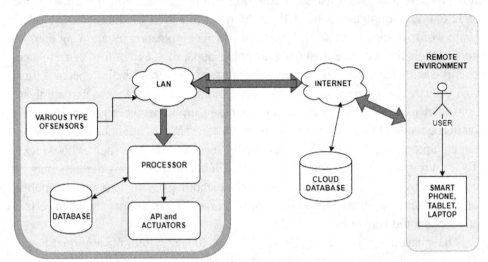

or may be a tablet. Internet is the core component of IoT to control these devices remotely.

History of Smart Home

Humans have tried to find out a way to make daily life easy and comfortable. The area of "smart home automation" depend on the technologies, its basic components are the internet and computer to process data in real-time. Scientists have been working last 35 years to make a fully smart home but still, they didn't receive the best output due to some difficulties such as internet issues, smart homes required aroused public internet, an aging society, and environmental situation. According to records, the first science fiction literature had come in the 1950's it portrayed the vision of homes, that were fully automatically controlled and monitored through the machine.

INTERNET OF THINGS (IOT)

The internet of things (IoT) is used to connect devices to the internet. Devices are physical equipment such as sensors and actuators, equipped with an interface known as telecommunication, a central process unit, a limit of storage, and application software. IoT established a connection among people, physical equipment (devices), and the internet. Manufacturing is one industry that has adopted cutting-edge technology like IoT, AI, facial recognition, deep learning, robots, and many more. Thanks to sensors that have been implanted and enable data transmission, factory robots are growing smarter. The robots' artificial intelligence systems allow them to learn from more recent data. This approach reduces time and costs while gradually improving the production process. IoT is collection of technologies, these technologies are:

1. **Sensors:** The sensor is an electronic device (remember that some sensors are not electronic), it takes input from the environment and converts it into data for further processing. This data is further processed either by the machine or human. For smart homes basically, we are using the electronic sensor. Electronic sensors are further categorized into two classes. The first one is an Analog sensor and the second one is a digital sensor. Analog sensors are used to convert sensor data into analog signals whereas digital sensors are used to convert what they are measuring in the input data to digital signals. To create a smart home both analog and digital sensors are used. Sensors could be coupled with any home appliance, such as an air conditioner, gas, light, doorbell, door lock, or any environmental device. With the assistance of sensor data, nowadays it is easy to monitor and control home functionality remotely by the Human.

2. **Software:** Software is a collection of instructions, which are a set of rules for doing work step by step. The computer uses instructions to perform each task. A collection of instructions is called software. The software is used to perform specific tasks. The software is application specific. No single software is used for multiple different applications. The software is in contact with the hardware, and the hardware is operated through the software.

3. **Connectivity:** Connections are used to connect systems or applications. Connection is a way to connect multiple applications or devices together. Connectivity is a way to connect your IoT device to a server or cloud. Data can be transferred from one device to another over the connection. Internet connection is a good example of a connection. If the two devices are not connected, they will not be able to share data.

4. **Location:** The location is the exact location of the person or object where the event occurred. Sensors are used to create intelligent applications. Get the exact position of the object using the GPS sensor. GPS is an abbreviation for Global Positioning System. GPS is the best sensor for getting accurate position with smart technology.

5. **Robotics:** Robotics is used to design intelligent machines that help to work automatically. It is used to support people. Robotics is an integrated field that includes mechanical engineering, electrical engineering, mechatronics, information engineering, computer engineering, mathematics, software engineering, control engineering and more. Robotics is used to create machines

that can replace humans and convey human behavior. Robots can be used for different purposes in different situations, such as fire area inspection. Many robots are used in dangerous situations such as radioactive material inspection, explosives detection, and bomb detection.

6. **Data Analytics:** Processing small units of data is easy, but large amounts of data are processed using algorithms such as machine algorithms. Data analysis includes extensive data analysis, AI and cognition, and edge analysis. Data analysis is used to examine datasets, draw trends based on trends, and draw conclusions. Data analysis is used to retrieve the information contained in a dataset. Specific hardware and software are used for data analysis.

INTERNET OF THINGS (IOT) FOR SMART HOME

IoT is a network of interconnected smart devices. Here smart device means any device, vehicle, or building that are embedded with electronics, sensor, actuators, software, or the internet. These devices are enough capable to collect and share data. IoT devices have been used to collect large amounts of data, including environment, weather, traffic, people, and more. All of these data collection elements are based on the accuracy of the analytical capabilities of this technology. The range of IoT sensors (Beretas et al., 2018) is not limited. Basically, it doesn't matter where and how you do it, depending on your needs. Sensors are used in almost every field, including: B. Sensors used in homes (Yun et al., 2013), organizations and agriculture. For example, sensors are used at home to help people. Sensors are used to monitor household items at home. IoT stands for Infrastructure of the Information society, and it was defined by IoT-GSI (Global Standards Initiative on Internet of Things) in 2013. As shown in Figure 3.

1. **Device connection:** IoT devices, IoT connectivity, embedded intelligence (Gunawan et al., 2017). When physical devices are connected together to share information through the Internet, it is known as connected devices. These devices are embedded in technologies such as processing chips, software, and hardware. Connected devices are a combination of various hardware such as sensors, computers and mobile phones. These devices can be controlled remotely via a smartphone, tablet, or computer system. These devices are remotely monitored and controlled using smart devices.

2. **Data Sensing:** Capture data, Sensors, and tags Storage. A sensor (Saraju et al., 2015) is an electronic machine, module, or device or subsystem that sends mutual signals to a server to detect, recognize, and further process events around the sensor. Sensors are used to capture events in the environment to

collect data. Capturing an event with a sensor is called a data capture. This is also known as sensor data acquisition.

3. **Communication:** Focus on access networks, cloud, edge, and data transport. IoT device communication is an infrastructure or system for exchanging data or information between devices connected to a system. All devices connected to the system can share data with each other. This system must be connected to the internet to communicate. The data collected by IoT devices (Farhan et al., 2018) is shared with devices connected via the Internet. Sensor data can be sent to the cloud globally or locally for analysis. These devices communicate with other connected devices to get information and data.

4. **Data analytics:** Big data analysis (Al-Ali et al., 2017a, Al-Ali et al., 2017b), AI, and cognitive, Analysis at the edge. A small unit of data processing is easy however a large amount data process using algorithms such as Machine Algorithm. Data analytics includes substantial data analysis, AI and cognitive, and the analysis at the edge. Data analysis is used to investigate datasets and draw trends based on trends to draw conclusions. Data analysis is used to retrieve the information contained in a dataset. Specific hardware and software are used for data analysis.

5. **Data value:** Analysis to action, APIs and processes, actionable intelligence. Data value is a way and it is used to take action on the data. Its perform analysis action, APIs and processes, and actionable intelligence also. The content used to enter records is called a data value. A database is a collection of various data fields such as numbers, names, and contacts. There are different types of IoT data, such as state data and automated data (Smita et al., 2017). Status data is structural data, line-based data, used to convey the status of connected devices. Automation data is collected from automated devices such as Automatic lighting was created.

6. **Human value:** Smart applications, Stakeholder benefits, Tangible benefits. Technology has become a part of human life, and the invention of the IoT has made human life easier. Just as human life has grown with technological capabilities, the IoT is growing day by day. With the help of IoT communications, exchanging data and interacting with devices has become easier. IoT works with security. Today, humans move online with IoT devices, not to exchange information. Technology becomes a part of human life, and the more inventions of technology make it easier.

The internet of things (IoT) is used to connect devices to the internet. Devices are physical equipment such as sensors and actuators, equipped with an interface known as telecommunication (Smita et al., 2017), a central process unit, a limit of storage, and application software. IoT established a connection among people,

Figure 3. IoT Work Process on Data

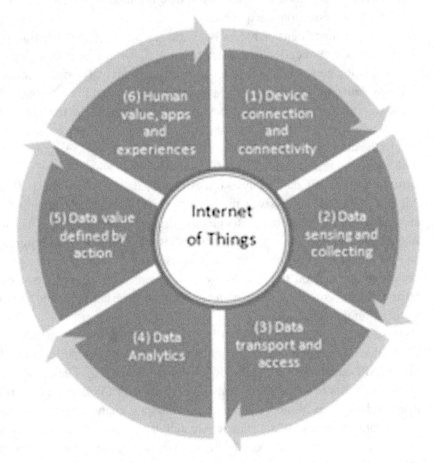

physical equipment (devices), and the internet. A core component of IoT is radio frequency identification (RFID), sensor and intelligence technology. RFID is a key point of networking. RFID (Malik & Tarar, 2021) processing and communication technologies are unique, they allow integration of the element to operate an integrated unit at the same time other devices can be added or removed easily. In the present scenario to enhance the bandwidth, it used JSON.

To create a smart home, the basic compounds which are required are IoT, cloud computing, AI, and the rule-based event listener. Each compound has its technologies to get composition. IoT is one of the core components of the smart home, IoT devices are a combination of internet connection, sensors, and software. IoT devices are remotely managed through the mobile application. A particular application is installed in smart mobile to control a particular smart device.

Figure 4. Block Diagram of Smart Home

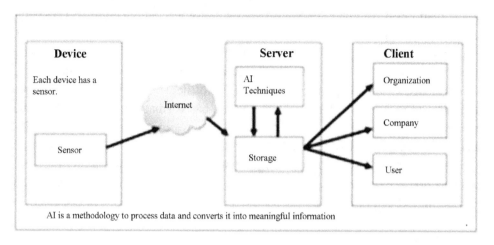

CLOUD COMPUTING AND ITS CONTRIBUTION TO IOT FOR SMART HOME

Cloud computing is a very important and useful technology to share resources and data over the internet in a virtual way. Cloud computing is a virtual service system to provides required resources and data through the internet. Cloud service (Malik & Tarar, 2021) could be accessed anywhere at any time without any time and location restrictions. The smart home is a collection of merged technologies such as IoT, cloud computing, and AI. Instead of this, these technologies are responsible to manage local and central computing with the optimized resource. Figure 4 is showing the architecture of cloud computing and IoT based smart home.

A computing task should be executed either on the IoT and smart home devices or outsourced to the cloud. Where to compute depends on the overhead tradeoffs, data availability, data dependency, amount of data transportation, communications dependency, and security considerations. On the one hand, the triple computing model involving the cloud, IoT, and smart home (Jadon et al., 2020), should minimize the entire system cost, usually with more focus on reducing resource consumption at home. On the other hand, an IoT and smart home computing service model should improve IoT users fulfill their demands when using cloud applications and address complex problems arising from the new IoT, smart home, and cloud service model. The idea of a "smart house" is one of pervasive computing that is progressively gaining popularity among people working in the tech industry. Users find it challenging to learn smart home (Li et al., 2022) due to the tremendous burden that intricate control (K., 2020) and a lot of data place on the local computer. This article suggests a smart house architecture based on cloud computing. Users may now get real-time information straight from their web browser, reducing the local effort. Additionally, a test platform to validate the cloud-based smart house architecture will be built. The proposed structure of the smart house is more practical, adaptable, effective, and economical, according to experimental findings. We have discovered via testing that cloud computing-based structures are accessible and have a greater range. Additionally, cloud computing offers a wealth of network resources and ensures user data protection. We have created a cloud-based smart house that boosts the adaptability and mobility of smart homes while lowering expenses. Technologies like cloud computing and the Internet of Things are evolving quickly. Both want to offer plenty of resources and a comfortable way of life. This article aims to suggest a smart house based on cloud computing by fusing smart home apps and cloud computing behavior patterns.

Integration of IoT and AI

In the current era, lots of Artificial Intelligence (AI) (Kashan Ali Shah & Mahmood, 2020) and Internet of Things (IoT) exertion have been created wonderfully, Out of these sometimes low latency with fast responses are demanded. Internet technologies have been developed in amazing view, these technologies are collaborating with service-oriented applications, AI and IoT to make the electronic device a smart device. Nowadays IoT with AI and cloud (Mayub et al., 2019) play an important role to enhance smart Engineering. AI and IoT technologies are capable to connect billions of intelligent machines to share information in the world.

Sensor

The sensor is an electronic device (Khanum & Shivakumar, 2019) (remember that some sensors are not electronic), it takes input from the environment and converts it into data for further processing. This data is further processed either by the machine or human. For smart homes basically, we are using the electronic sensor. Electronic sensors are further categorized into two classes. The first one is an Analog sensor and the second one is a digital sensor. Analog sensors are used to convert sensor data into analog signals whereas digital sensors are used to convert what they are measuring in the input data to digital signals. To create a smart home both analog and digital sensors are used. Sensors could be coupled with any home appliance, such as an air conditioner, gas, light, doorbell, door lock, or any environmental device. With the assistance of sensor data, nowadays it is easy to monitor and control home (Kumar et al., 2021) functionality remotely by the Human.

Artificial Intelligence

Artificial intelligence is a part of computer science which is emphasized to create intelligent machines or devices like a human. Human is capable of deciding according to real scenario situation that normal machine cannot do. Now AI works on artificial intelligence machines that can react like a human, these machines can decide on a human without any external instruction. Integration of AI and IoT (Dipali et al., 2016) work to make smart homes make human life easy. Alexa, Siri, and Google Assistant are AI products, and all are used to control smart home devices through voice.

How is AI used in IoT?

The corporate environment has altered since IoT was adopted. Data from many sources is prominently captured via IoT. Data wrapping has been applied to several IoT devices, though. To gather data, several IoT devices are employed simultaneously. Data was gathered, prepared, and examined to get a certain result. AI and IoT Both technologies are utilized to obtain the smart device, but they operate in different ways. For example, the Internet of Things (IoT) (Risteska Stojkoska & Trivodalicv, 2017) gives data through sensors, whilst artificial intelligence is used to unlock replies. Sensor-delivered data is analyzed by AI, and the end result is a business decision that is informed. The following are made agile with IoT and AI:

- They are employed to organize, examine, and extract pertinent data from the data.
- Control, examine, and get valuable insights from data.
- Ensure quick and precise analysis
- Strike a balance between the needs for centralized and localized intelligence
- Strike a balance between personalization, privacy, and secrecy.
- Continue to protect against cyber attacks

AI is utilized in IoT to evaluate the ongoing data streams and identify patterns that aren't readily apparent from simple gauges. AI and machine learning might potentially anticipate operating conditions and pinpoint which variables need to be adjusted for the best outcomes. Therefore, information about which operations may be changed to boost efficiency and which ones are unnecessary and time-consuming is provided by intelligent IoT devices. IoT devices (Romeo, 2019) range from pricey sensors to high-end PCs and mobile phones. However, the IoT ecosystem that is most often employed consists of low-cost sensors that generate enormous volumes of data. An IoT ecosystem powered (Al-Hassan et al., 2018) by AI assesses and summarizes the data from one device before delivering it to other devices. As a result, it makes enormous volumes of data manageable and makes it possible for several IoT devices to communicate. This is what we mean by scalability.

It would need even more investment in new technologies to fully utilize the promise of IoT devices in the future. AI and IoT integration have provided new methods for organizations, economies, and sectors to operate. AI has the capacity to build intelligent machines that can replicate player intelligence and aid in decision-making with the least amount of human input. Humanity gains a lot from the combination of these two technologies. IoT is concerned with internet-connected devices, whereas AI creates computers that can learn from data and their own prior experiences.

ADVANTAGE OF SMART HOME

The smart home has a huge number of advantages:

- **Lock and Unlock Home Door easily:** With the help of smart home devices (A. Mani et al., 2018), you can lock or unlock your home door remotely through the controller. If any unauthorized user tries to unlock your home door a notification or alert comes on your smartphone.
- **Smart Energy Consumption is used to Save Energy:** With the help of a smart home, you can off lights, AC, and Coffee maker remotely. If a person forgets to turn off AC while it going to the office it can off from the office through the smartphone it becomes possible through smart home technologies.
- **Used to maintenance and service:** Smart home (Yan, 2016) service makes human life easy. All things at home can manage easily.
- **User can easily customize things as per its convenience:** Users can manage things easily and can control them remotely.
- **Easy to use smart home technologies:** Smart home technologies are easy to control through the controller.

DISADVANTAGE OF SMART HOME

We are aware of the advantage of a smart home but it also has some disadvantages.

- **Cost:** One of the biggest issues with a smart home is cost. Most of the companies provide smart home services but they are expensive too. Everyone cannot afford a smart home. Many hardware, software, and internet connection are required.

- **Maintenance Cost:** Smart device maintenance (Yar et al., 2021) cost is also high. You have to repair smart devices from time to time if it's required.
- **Internet Dependency:** Without an internet connection, smart devices cannot be operated. To control or access the functionality of smart devices, high bandwidth internet connection is required.
- **Dependency on Controller:** To operate smart devices, one controller machine is required it may be a smartphone, tablet, or PC.
- **Dependency on Professionals:** To operate smart devices, the user must know of them. Everyone cannot set up a smart home directly, a third party (means an organization) that is provide smart home service is required to set up a smart home.

CONCLUSION

The smart home is good in terms of time and efficiency. Human life becomes easy through a smart home; one of the best things about a smart home is its security system. The smart home is more secure as compared to a normal home. One of the advantages of smart is remotely controlling all the devices. Now house owner is independent to control physical devices not depend on the location. Instead of these features few points also be considered as a disadvantage of the smart home that is, Maintenance cost is too high, the internet connection must be required, and the internet connection should not be poor. Smart home cost is high and they cannot be affordable by normal people. Hence most people want a smart home because it gives comfort and time managing concepts. Many industries provide services to build your home as a smart home to make human life easy.

REFERENCES

A, J., Nagarajan, R., Satheeshkumar, K., Ajithkumar, N., Gopinath, P., & Ranjithkumar, S. (2017). Intelligent Smart Home Automation and Security System Using Arduino and Wi-fi. *International Journal of Engineering And Computer Science*. doi:10.18535/ijecs/v6i3.53

Al-Ali, A., Zualkernan, I. A., Rashid, M., Gupta, R., & Alikarar, M. (2017). A smart home energy management system using IoT and big data analytics approach. *IEEE Transactions on Consumer Electronics*, *63*(4), 426–434. doi:10.1109/TCE.2017.015014

Al-Hassan, E., Shareef, H., Islam, M. M., Wahyudie, A., & Abdrabou, A. A. (2018). Improved Smart Power Socket for Monitoring and Controlling Electrical Home Appliances. *IEEE Access: Practical Innovations, Open Solutions*, 6, 49292–49305. doi:10.1109/ACCESS.2018.2868788

Alaa, M., Zaidan, A., Zaidan, B., Talal, M., & Kiah, M. (2017). A review of smart home applications based on Internet of Things. *Journal of Network and Computer Applications*, 97, 48–65. doi:10.1016/j.jnca.2017.08.017

Beretas, C. (2019). Internet of Things (IOT) is Smart Homes and the Risks. *Journal of Current Engineering and Technology*, 1(3), 1–2. doi:10.36266/JCET/115

Ghorpade, D. D., & Patki, A. M. (2016). A Review on IOT Based Smart Home Automation Using Renewable Energy Sources. *International Journal of Scientific Research*, 5(5), 2000–2001. doi:10.21275/v5i5.NOV163878

Gunawan, T. S., Yaldi, I. R. H., Kartiwi, M., Ismail, N., Za'bah, N. F., Mansor, H., & Nordin, A. N. (2017). Prototype Design of Smart Home System using Internet of Things. *Indonesian Journal of Electrical Engineering and Computer Science*, 7(1), 107. doi:10.11591/ijeecs.v7.i1.pp107-115

Jadon, S., Choudhary, A., Saini, H., Dua, U., Sharma, N., & Kaushik, I. (2020). Comfy Smart Home using IoT. SSRN *Electronic Journal*. doi:10.2139/ssrn.3565908

K., D. S. (2020). IoT based Smart Energy Theft Detection System in Smart Home. *Journal of Advanced Research in Dynamical and Control Systems, 12*(SP8), 605–613. doi:10.5373/JARDCS/V12SP8/20202561

Kashan Ali Shah, S., & Mahmood, W. (2020). Smart Home Automation Using IOT and its Low Cost Implementation. *International Journal of Engineering and Manufacturing*, 10(5), 28–36. doi:10.5815/ijem.2020.05.03

Khanum, A., & Shivakumar, R. (2019). An enhanced security alert system for smart home using IOT. *Indonesian Journal of Electrical Engineering and Computer Science*, 13(1), 27. doi:10.11591/ijeecs.v13.i1.pp27-34

Kumar, P. P., Prasanth, P. P., Hemalatha, P., & Kulakarni, K. J. (2021). A Framework for Fully Automated Home using IoT Reliable Protocol Stack and Smart Gateway. *International Journal of Robotics and Automation Technology*, 7, 56–62. doi:10.31875/2409-9694.2020.07.7

Li, A., Bodanese, E., Poslad, S., Hou, T., Wu, K., & Luo, F. (2022). A Trajectory-based Gesture Recognition in Smart Homes based on the Ultra-Wideband Communication System. *IEEE Internet of Things Journal*, *1*, 1. Advance online publication. doi:10.1109/JIOT.2022.3185084

Mahindrakar, S., & Biradar, R. K. (2017a). Internet of Things: Smart Home Automation System using Raspberry Pi. *International Journal of Scientific Research*, *6*(1), 901–905. doi:10.21275/ART20164204

Malik, I., & Tarar, S. (2021). Cloud-Based Smart City Using Internet of Things. *Integration and Implementation of the Internet of Things Through Cloud Computing*, 133–154. doi:10.4018/978-1-7998-6981-8.ch007

Mani, A., & Charan, R. (2018). A survey on IoT based system for smart home automation and theft control. *International Journal of Modern Trends in Engineering & Research*, *5*(2), 70–74. doi:10.21884/IJMTER.2018.5038.VVILQ

Mayub, A., Fahmizal, F., Shidiq, M., Oktiawati, U. Y., & Rosyid, N. R. (2019). Implementation smart home using internet of things. *TELKOMNIKA*, *17*(6), 3126. doi:10.12928/telkomnika.v17i6.11722

Mohanty, S. P. (2015). IEEE ISCE. *IEEE Consumer Electronics Magazine*, *4*(1), 111. doi:10.1109/MCE.2014.2361003

Rafi, F. A. (2018). Implementation of smart home automation using raspberry pi. *International Journal of Recent Trends in Engineering and Research*, 122–125. Advance online publication. doi:10.23883/IJRTER.CONF.02180328.019.HDV3S

Risteska Stojkoska, B. L., & Trivodaliev, K. V. (2017). A review of Internet of Things for smart home: Challenges and solutions. *Journal of Cleaner Production*, *140*, 1454–1464. doi:10.1016/j.jclepro.2016.10.006

Romeo, M. S. S. (2019). Intrusion Detection System (IDS) in Internet of Things (IoT) Devices for Smart Home. *International Journal of Psychosocial Rehabilitation*, *23*(4), 1217–1227. doi:10.37200/IJPR/V23I4/PR190448

Yan, Z. (2016). Complex Systems Smart Home Security Studies based Big Data Analytics. *International Journal of Smart Home*, *10*(6), 41–52. doi:10.14257/ijsh.2016.10.6.05

Yar, H., Imran, A. S., Khan, Z. A., Sajjad, M., & Kastrati, Z. (2021). Towards Smart Home Automation Using IoT-Enabled Edge-Computing Paradigm. *Sensors (Basel)*, *21*(14), 4932. doi:10.339021144932 PMID:34300671

Yun, J., Li, X., Wei Wang, S. L., & Li, L. (2013). Research and Development of LBS-based Ethnic Cultural Information Resources Smart Recommendation System in Mobile Internet. *Information Technology Journal, 12*(21), 6440–6448. doi:10.3923/itj.2013.6440.6448

Chapter 9
Smart Bin:
An Approach to Design Smart Waste Management for Smart Cities

Divya Upadhyay

https://orcid.org/0000-0003-3664-7415
ABES Engineering College, Ghaziabad, India

Ayushi Agarwal
ABES Engineering College, Ghaziabad, India

ABSTRACT

The 21st century is the era of the digital world and advanced technologies. This chapter contributes to the Swachh Bharat mission by presenting the concept of smart bin using IoT. The Smart bin presented in this chapter is GPS-enabled and comprises sensors and a camera. A prototype for the proposed model is analysed, and network architecture is designed to communicate the critical information. The proposed system will update the status and condition of the bin to the nearest authority to improve the city's pollution and cleanliness. The prototype is deployed using a microcontroller Raspberry Pi and Google Maps to obtain the bins' real-time location. IoT fill-level sensors will help the garbage carrying truck in identify the nearest empty container without wasting time and resources. Google Maps will help in sensing the optimised routes to the drivers. The microcontroller will be used to integrate the different devices and cameras to provide real-time bin collection, overflowing/under flowing state and tracking information, and suggestions and notifications for effective disposal.

DOI: 10.4018/978-1-6684-4991-2.ch009

INTRODUCTION

India's population of around 1.2 billion generates about 0.1 million tons of waste daily. The local municipal bodies do waste management in the country, mostly found to dump the trash on open ground. Seldom is it burnt and creates environmental pollution. Swachh Bharat Mission, started by our honorable Prime Minister in October 2014, is motivated to properly dispose of trash and improve the city's cleanliness. The advancements in smart cities across the globe are because of urbanization, increased industries, population, governments, and technological advancements (Thakker & Narayanamoorthi, 2015). Therefore, a considerable sum of money is invested by governments in establishing smart cities. The concept of smart bins is going to influence human lives. Therefore, a need for efficient trash management is mandatory.

Smart-Bins concept comes into the picture by providing an environment-friendly and sustainable solution to the problem of waste-management. Smart-Bins have been known to manage waste and trash efficiently. Its cameras and sensors help sends real-time signals and monitors the trash. The usage of Smart-Bins can be accommodated at two levels that are: home-based and on the streets. The use of Smart-Bins at home-based is mainly concerned with processing waste at an introductory level. The Smart-Bins at home-based are primitive and use sensors and cameras. Whenever the trash reaches the threshold level, it sends messages and notifications to the user. Some Smart-Bins also process the trash to produce a helpful substance that can be used for providing nutrition to the plants. Smart-Bins have a more efficient storage capacity than traditional bins of as much as 80%.

The Smart-Bins used in the streets have a higher storage capacity than those used at home-based. The municipal authorities use these Smart-Bins with features like real-time location-monitoring, cameras and sensors installed. The signals are sent to municipal authorities whenever the trash reaches a threshold limit. They utilize the local Wi-Fi for sending the signals to concerned authorities. As a result, it reduces the effort required for the authorities to collect the trash. The trash collectors don't have to visit the bins every day, minimizing the trash collector's efforts and saving time. These bins are environmentally friendly and reduce the harmful gases and lousy odor released from the trash. Therefore, this ideal will help reduce air pollution in the area and help in the prevention of the spreading of viruses caused by unmanaged waste (Kumari, 2018).

The main objectives of this proposed work are:

i. To design an IoT-aware trash-monitoring system that will leverage the features of various sensors to analyze the Smart-Bin's status.

Figure 1. Basic overview of the Smart-Bin system

ii. The proposed system will also efficiently review and monitor the collected and actual data of the ongoing process mechanism.

iii. To design and implement a full-fledged architecture for the intelligent bin working and data transmission between the system and the user.

iv. The proposed system reviews the temperature of the trash monitoring system by the data collected through sensors.

Figure 2. Concept of Smart-Bin

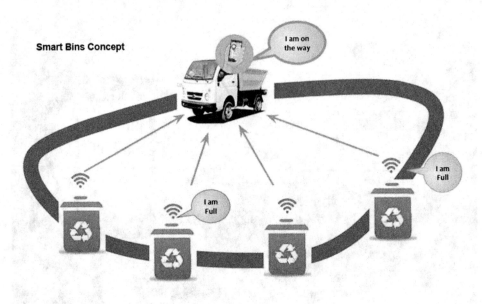

SCOPE

"Smart-Bin is used to identify the status of empty or full waste bins, adjust the waste collection schedule accordingly and save costs. The real-time waste management system uses sensors to check if bins are full. Using the proposed system, the person concerned can access the information of all smart bins from anywhere at any time. It will report the status of each bin in real-time, so the relevant authority can send a vehicle to collect waste only when the bin is full. By implementing this system, optimization of resources, cost reduction, and practical use of smart bins can be achieved.

Smart-Bin explains bins' scope of work in managing the entire city's waste collection system. The sensor-enabled Smart-Bins connected through the cellular-network generate a large amount of data, which is investigated and visualized in a real-time to gain insight perceptions into the status of waste and trash around the city. The scope for the future work of this system is it can be implemented in various places like government offices; Bus stops, Railway site-stations, Local Markets, Metro site-stations, Localized dustbins in each regional sector, Colleges, Schools, Hospitals, Restaurants and many more. The figure below will display the backside-street information stickers, facilities like CCTV, Wi-Fi Hotspot, LED Signboard, and the Bio Enable Smart-Bin sensor inside the dustbin. Below are the listed features applicable for the Multipurpose Smart Street Bin:

- Smart-Bin with various sensors
- LED-based Display for public information
- Street/Direction Information Boards
- Wi-Fi Hotspot
- CCTV-based Camera Monitoring"

RELATED WORK

Smart-Bins are the next-generation technologies changing conventional trash and garbage collection methods. The advantages of these smart technology have prompted state municipal authorities and various households to adopt and implement them. The efficient collection and its management contribute to its adoption. The underlying technology behind the Smart-Bins is the advancements of the Internet of Things (IoT) and sensors; it allows for the real-time monitoring of the trash levels and informs the user/authorities about when the trash should be emptied. It is capable of monitoring and analyzing the contents of the bin. In recent times, some of the valuable aspects of intelligent and smart bins have been converting the trash into manure and unprocessed fertilizer that can be utilized for the plants.

Apart from that, Smart-Bins are also utilizing solar-power for operation, resulting in less dependency on electrical energy. The most significant advantage being offered is the efficient utilization of space, allowing it to use 80% less space than the conventional approach. The operation and functions of the Smart-Bins also ensure that the trash's odor is reduced and there is less emission of greenhouse gases. In recent works, the IoT has played an important role in improving human life by designing IoT-based adequate waste collection systems with Smart-Bins (Srisabarimani et al., 2022). A waste classification system was designed and named Waste Net, based on the convolutional networks.

A collaborative plastic waste management infrastructure has been designed to help seniors and the elderly. According to new research (John et al., 2021), plastic waste in the current scenario occupies a good space in every house that needs to be managed.

In another continuation of the research (John et al., 2021), IoT technology is used to create a monitoring system in smart cities.

Plastic waste is one of the serious causes of pollution. Plastic waste management needs more attention as many types of research have been explored. One of them, Sidhu et al. (2021), worked to create a collaborative application to help manage household plastic waste through Smart-Bins. It is necessary for households because most daily plastic waste comes from households.

According to new research, it also became clear that smart cities require more waste management, and these works were done for Smart-Bins that would monitor waste in smart cities using IoT systems (Ravi et al., 2021).

The Long-Range Wide Area Network (LoRaWAN) is so fast in the tech industry that it's creating an IoT-based waste management system. This idea can be explored in the paper (Baldo et al., 2021). Plastic waste is one of the most evaluated causes of the increase in pollution. Thus, to solve plastic waste management, a set of sensors such as IR sensor, ultra-sonic sensor and capacitive sensor are used to sense the presence of waste, measure the level of waste and identify the presence of plastic. Foul smell and gases evolved from waste is another problem humanity faces and struggles with. To simplify, a sensor system for detecting waste gases has been proposed (Sanger et al., 2021).

In the next procedure, the problem was that people sometimes faced many obstacles when moving garbage from the dustbins to the main garbage depot in the municipality. Even a member of the municipality makes a great effort. One of the researchers proposed to overcome this system by creating a scenario to reduce the travel distance by up to 30.76% (Haque et al., 2020).

Manufacturers of Smart-Bins use RFID tags to manufacture Smart-Bins (Praveen et al., 2018). Smart-Bins has reduced the effort involved in their collection and helped authorities allocate their resources more efficiently and promptly. The use of spectroscopy to separate biodegradable and non-biodegradable waste was proposed by (Thakker & Narayanamoorthi, 2015). The Global System for Mobile Communications (GSM MODULE) and ARM (Acorn RISC Machine) 7 were designed for smart waste management using quality of life as a measure index. Smart-Bins will enhance warehouse management, as suggested by Suleman Ahmed (Ramson & Moni, 2017). A paper by (Ali et al., 2012) suggested the wider use of Wireless Sensor Networks (WSNs) to monitor Smart-Bins and empty trash levels using central monitoring systems and wireless links.

The Smart-Bin concept is a very noble idea these days. Smart-Bin is built on the Arduino board platform. It is connected to a GSM MODULE and Wi-Fi module, and the basket is equipped with an ultra-sonic sensor (HC-SR04) and a temperature sensor (DHT-11). A working Smart-Bin is shown in Figure.3.

It monitors the trash level in the bin and alerts the user about the bin's state. There are various components used in the system:

- Arduino Board
- Ultra-sonic Sensor
- GSM MODULE
- Temperature Sensor
- Wi-Fi module

Figure 3. Pictorial representation of Smart-Bin

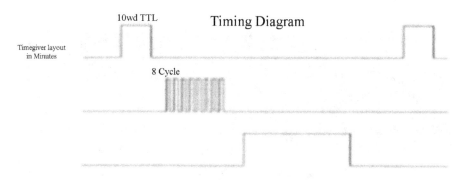

- Connecting wires

Arduino

Arduino is an open-source computer hardware and software company, project and user community that designs and manufactures microcontroller kits for building digital devices and interactive objects that can sense and control things in the physical world. The project's products are distributed as open-course hardware and software, licensed under the GNU Lesser General Public License (LGPL) or GNU General Public License (GPL), which allows anyone to make Arduino boards and distribute software. Arduino boards are commercially available in the pre-assembled form or as DIY kits. Arduino board designs use different microprocessors and controllers. The UN can drill without fear of doing anything wrong; at worst, the chip can be replaced for a few dollars and start over.

"Arduino contains several different parts and interfaces on a single circuit board. The design has changed over the years, and some variants include additional features. The Arduino board includes several pins connect to different components with Arduino. Arduino board in detail is presented:

GSM MODULE

Global-System for Mobile communication (GSM MODULE) is a digital mobile network widely used by mobile-phone users in Europe and other parts of the world. GSM MODULE uses a variation of Time-Division Multiple-Access (TDMA) and is the most widely used digital wireless telephony technology. Other technologies are TDMA, GSM MODULE, and Code-Division Mutliple-Access (CDMA). GSM MODULE digitizes and compresses the signal data and sends it over a channel with two other streams of user data, each in its time slot. It operates in the 900 MHZ

Figure 4a. Arduino Board

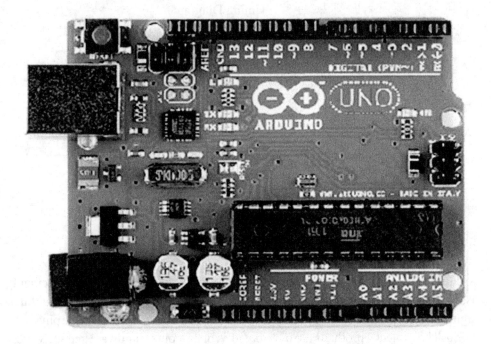

or 1800 MHZ frequency band. GSM MODULE-based mobile services were first launched in Finland in 1991.

A GSM MODULE network consists of many functional units. These functions and interfaces are explained in this section. Figure.5, given below, is a simple pictorial view of the GSM MODULE architecture:

The additional components of the GSM MODULE architecture comprise databases and messaging systems functions: Homebased Location-Register (HLR), Equipment Identity-Register (EIR), Visitor Location-Register (VLR), Authentication-Cener (AuC), SMS-Serving Center (SMS SC), Transcoder and Adaption Unit (TRAU), Chargeback Center (CBC), and Gateway MSC (GMSC).

Wi-Fi Module

ESP-8266 is a Wi-Fi enabled system-on-chip (SoC) module developed and designed by the Esp-ressif system. It is used to create embedded IoT applications. Figure 6 depicts the view of the Wi-Fi Module.

ESP-8266 comes with capabilities of

Figure 4b. Arduino Board Description

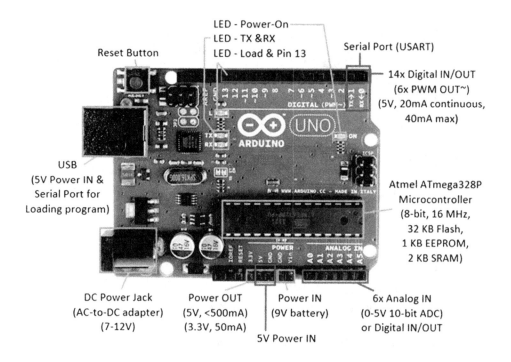

Figure 5a. GSM Architecture for Smart Bin

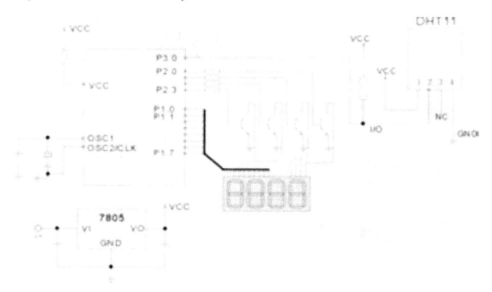

Figure 5b. GSM MODULE Architecture for Smart-Bin

Smart Bins Concept

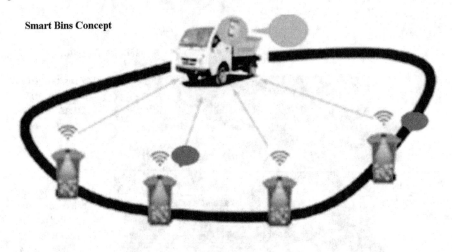

- 2.4 GHz Wi-Fi (802.11 b/g/n, supporting WPA/WPA2),
- Pulse-width modulation (PWM)
- Analogue-to-digital conversation (10-bit ADC)
- Serial Peripheral Interface (SPI) serial communication protocol,
- I2S (Inter-IC Sound) interfaces with DMA (Direct Memory Access (sharing pins with (GPIO),

Figure 5c. GSM MODULE Architecture Service area

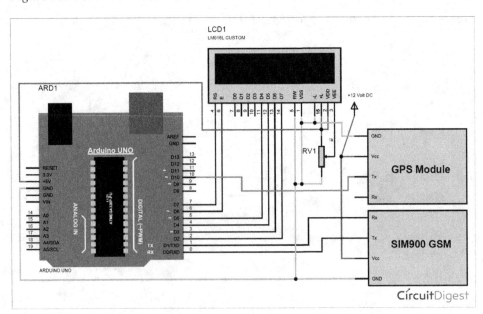

Figure 6. ESP-8266-01 Wi-Fi Module

ESP8266-01 WIFI Module

- UART (on dedicated pins, plus transmit-only UART can be enabled on GPIO2),
- General-purpose input/output (16 GPIO)
- Inter-Integrated Circuit (I2C) serial communication protocol,

The ESP-8266 Module is a low-cost and stand-alone wireless transceiver. It can be deployed to develop IoT endpoints.

The ESP-8266 Module communicates with a micro-controller that uses a set of AT commands, The microcontroller communicates with the ESP8266-01 Module using a UART with specified baud rate.

Ultra-Sonic Sensor

The HC-SR 04 ultra-sonic sensor uses the SONAR waves to determine the distance of an object, just like bats do. Ultra-sonic sensors can avoid collision for the robot and are often relocated not very fast. Ultra-sonics are so widely used that they can be reliably implemented in grain bin sensing applications, water level sensing, drone applications and car sensing in a local restaurant or bank. Ultra-sonic rangefinders ae commonly used as collision detection devises.

Here's a list of some of the HC-SR04 ultra-sonic sensor features and specs:

§ Power Supply:+5V DC

Figure 7. Ultra-sonic Sensor

§ Ranging Distance: 2cm – 400 cm/1^2 – 13ft
§ Effectual Angle: <15°
§ Trigger Input Pulse width: 10uS
§ Working Current: 15mA
§ Resolution: 0.3 cm
§ Quiescent Current: <2mA
§ Dimension: 45mm x 20mm x 15mm
§ Measuring Angle: 30 degree

This diagram shows how the ultra-sonic sensor works when detecting an obstacle and measuring the distance between the block and itself.

It is a table that compares all the previous algorithms and research on the same technology. The contrast between various methods and the types of sensors and other tools used. The proposed work has analyzed all the required techniques and adopted a new algorithm for the current working system. The common factor between each method is the IoT techniques deployed. The proposed system will deploy the distance measurement and send the data to the associated app for ease of use for the concerned users. Each technique has its advantages.

Figure 8. shows the reflected wave echo is generated.

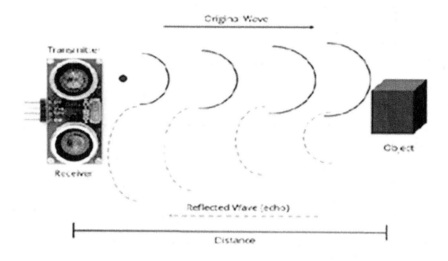

Smart-Bin Design and Analysis

Planning

Initial planning for the system's successful deployment was carried out in the phase. Identification of needs for the Smart-Bin concept was studied by evaluating and analyzing issues with the traditional waste management system. The list of requirements was made based on the issues with the traditional and normal bins. An economical, technical, stable and organizational approved feasibility study was carried out. It helped in determining whether the system is practicable by waste collection authority.

Requirement Gathering and Analysis

Gathering requirements is an important step. It was collected for primary and secondary data to analyze traditional waste management practices and systems. It helped in identifying research for similar products in the industry. This established the motivation for the basic concept of a new smart and intelligent system. A review of the waste management department and only surveys and questionnaires were used as primary data collection tools. In contrast, a thorough literature review was used to collect secondary data. The collected data was rigorously analyzed to identify and recognize the users and their basic requirements for the new system.

Table 1. Detail information for the previous works on Smart-Bins.

S. No	SENSOR	TECHNIQUES	APPLICATION	OUTPUT	PAPER
1.	1. Temperature (DHT-11) 2. Distance sensor (HY-SR05)	Model-based on: 1. Data Gathering Layer 2. Data Processing Layer 3. Data Demonstration Layer	It can be used as a sensor node for an integrated waste management system that runs in cloud system or fog computing system.	The model proposes a prototype model which can be deployed and implemented as a monitoring and notification system the municipal operators can use to realize the tractability of the waste sources.	Kumari, P. (2018, January 23).
2.	1. RFID, 2. Photoelectric sensor, 3. IR sensor, 4. weight sensor	The model is based on: 1. Data collected by RFID card readers and sensors		This paper concludes that RFID based bin is better than the IoT technique	Maheshwaran, K. (2018, April 24).
3.	1. Ultra-sonic sensor (HC-SR04) 2. Motion sensor (HC-SR501)	The model is based on: 1. An Arduino Nano board 2. Ultra-sonic sensor to monitor the fullness level of the container 3. Give SMS alerts using a GSM MODULE.		This model successfully achieved the parameters required outdoors as well as indoors.	Samann, F. E. F. (2017, August 30).
4.	1. Servo Motor	The model is based on: 1. Arduino technology 2. An RFID tag communicating 3. Servo motors. 4. Works on an automated bit technique	This model is an energy-efficient technique that uses power-down mode on the Arduino microcontroller and is applied to a recycling bin system.	The model eliminates the unnecessary wake-up of the microcontroller, reducing its energy consumption. The battery life of the autonomous device is increased by nearly 40.9%, which contributes to reducing maintenance downtime.	Suddul, G. (2018).
5.	1. PH sensor, 2. Infra-Red sensor 3. Ultra-sonic sensor	The model is based on: 1. GSM MODULE, RFID, 2. UART 3. Sensor communication. 4. To enhance waste collection efficiently using 5. RFID technology 6. Sensor systems.		Model is used for clearing trash and efficiently managing the overflow of waste. This system can also avoid fire accidents in trash cans with the help of a fire sensor. This will intimate or send SMS to the authorized person through GSM MODULE.	Amruta, B.P., Madhuri, S., Pallavi, P., & Satish, S. (2018).
6.	1. Ultra Violet Sensors	This model is based on: 1. Arduino Uno, 2. UV sensors, 3. GSM MODULE 4. Display unit. The test cases were chosen based on: 1. Realistic possibilities of offering the waste collection service depending on the filled bin ratio.	The model can be used in the smart city for proper waste management, especially in the newly planned Indian smart cities. Implementing it in major cities will raise new challenges.	The model prototype was used to verify the working capabilities. The experiments revealed the work providing validations in test cases and status identification.	Bano, A. (2020, December 29).
7.	1. Ultra-sonic Sensors	This paper consists of: 1. An ARM controller, 2. Ultra-sonic sensors, 3. A WIFI module 4. A website for interaction.	This work introduces the design and development of a smart green environment of the trash monitoring system by measuring the trash level in real-time and alerting the municipality where the bin is never full based on the types of trash.	The system calculates the distance and monitors trash through the Module. The proposed system will reduce the cost and workforce and indirectly reduce the traffic.	Mustafa, M. R. (2017).
8.	1. PIR sensors, 2. Ultra-sonic sensors	This model presents: 1. An IoT innovation project of a smart waste bin 2. A real-time monitoring system that integrates multiple technologies such as a. solar systems, sensors b. Wireless communication technologies	This model provides efficiency in trash monitoring and cost reduction.	The Smart-Bin works only on an internet connection. The desired performance and efficiency of the system are achieved. The Wi-Fi module provides an easy internet connection for the Arduino board.	Wijaya, A. S. (2017, August 1).
9.	1. Ultra-sonic sensors, 2. Humidity sensors 3. Temperature sensor	This model consists of various sensors: 1. ZigBee-PRO alliance communication module 2. Castalia simulator etc. 3. It works on the principle of the real-time monitoring system	The proposed system can be used by the waste management authority to help optimization of waste collection routes by using real-time waste fill level data. Thus the system can reduce pollutant emissions and operation costs.	The model presents a system that collects and monitors bin information as soon as someone throws waste inside a bin. The functions of the bin are to acquire various data about bin conditions, make some initial measurements on the bin level and then send the data to the control site-station.	Mamun, M.A., Hannan, M.A., & Hussain, A. (2014).
10.	1. IR-Sensors, 2. Ultra-sonic Sensors 3. Capacitive sensors	This model combines: 1. Different sensors 2. IoT technology for plastic waste segregation from the other waste as it is the most influential factor in causing pollution.	This model is designed to segregate waste and monitor real-time updates of dustbins using IoT.	The proposed system is well efficient in managing the defined task.	Srisabarimani, K., Arthi, R., Soundarya, S., Akshaya, A., Mahitha, M., Sudheer, S. (2022).
11.		This model is designed with: 1. Efficient mapping sensors to reduce the trash collector travelling in locating the trash bin location.	This model is efficient in decreasing the trash collector travel distance, and multiple sensors have made the system efficient.	This model resulted in a 30.76% decrease in the travel distance when tested for a specific pair of locations.	Haque, K. F., & Et. Al. (2020, June 1).

Figure 9. The Architectural flowchart for Smart-Bin.

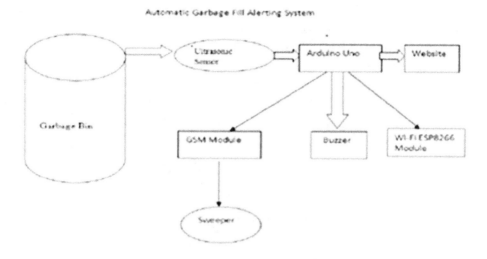

Design

The basic designs for the propsed Smart-Bin wie designed to determine the exact operations required from IoT devices. Design illustrations, a diagram of the Smart-Bins computer hardware infrasturcture, and desktop and Android applications interface designs were outlined. A hardware programming IDE, Android Development Kit, Desktop Application Platform, and database servers were decided upon. This review provided satisfactoy results showing that the project is feasible at an economic and opertaiotnal level.

Smart-Bin Architecture

This project aims to design a reliable and accurate IoT-based Smart-Bin for designing smart and intelligent cities. The system visualizes in real time by collecting data related to waste management and providing it to the control center. The collected data is analyzed at the decision support layer and alert, or warning messages are sent in case of emergency. Giant 1 illustrates the architecture for deploying sensors and Internet of Things. It consists of various sensors for monitoring. Data collected in real time is transmitted to the IoT cloud storage and further action.

The proposed Smart-Bin system will explain the complete hardware requirements and the connection between hardware and data. Figure 1 shows a basic overview of the system. When the sensor detects the status as full, the system will notify the appropriate user that the tank is full. The second part of the system sends an alert message to the control unit of the waste tracking system to collect the waste from the relevant Smart-Bin system. It will return to normal=empty once the medics have emptied it. The LED will flash until the bin is empty. The memory card records all usage and fullness alerts for later analysis. In addition, a motion sensor will be used to detect a usage event, which will play a thank you audio message stored on the memory card using a speaker to encourage the bin user.

Figure 10. Flowchart of the proposed model

Analysis of System

Certain experiments were performed on the developed system, which provided different results. But even before setting up the whole system, it is necessary to select the correct items to construct the required Smart-Bin. The perfect selection starts with analyzing the characteristics of the sensors required to be deployed. Two types of sensors were used in this work:

I.   Ultra-Sonic Sensor
II.   Temperature Sensor

Two sensors of each type are considered for comparison, which are used frequently in the market.

1)   Taking into consideration that there are four types of ultra-sonic sensors present

    a.   proximity sensor
    b.   two-point proximity switches
    c.   retro reflective sensor
    d.   beam sensor

The project requires the usage of distance-based calculation in the system, so two ultra-sonic sensors are shortlisted, HC-SR04 and HC-SR05. A comparison between the two was made.

Module of HC-SR04

Ultra-sonic ranging module HC-SR04 provides a 2cm-400cm non-contact measurement function; the ranging accuracy can reach 3mm. The modules include ultra-sonic transmitters, receivers and control circuits. The basic principle of work:

(2) Using IO trigger for at least 10us high-level signal,
(3) The module automatically sends eight 40 kHz and detects whether there is a pulse signal
(4) If the signal goes through a high level, high output IO duration is the time from sending ultra-sonic to returning.

Test distance = (high level time × velocity of sound (340M/S) /2 (1)
The wire directly connects as follows:

- 5V Supply
- Trigger Pulse Input
- Echo Pulse Output
- 0V Ground

Timing Diagram

The Timing diagram is shown below. A short 10uS pulse is supplied to the trigger input to start the ranging, and then the Module will send out an eight-cycle burst of ultra-sound at 40 kHz and raise its Echo. The Echo is a distance object that is pulse width and range in proportion. The range is calculated through the time interval between sending the trigger signal and receiving the echo signal.

Using an over 60ms measurement cycle is suggested to prevent trigger signals to the echo signal.

The timing diagram of HC-SR04 is shown in Figure.11.

Module of HC-SR05

It is also used to measure the distance between the obstacle and the sensor. With slight variations to SR04, it works similarly.

Figure 11. The timing diagram of HC-SR04

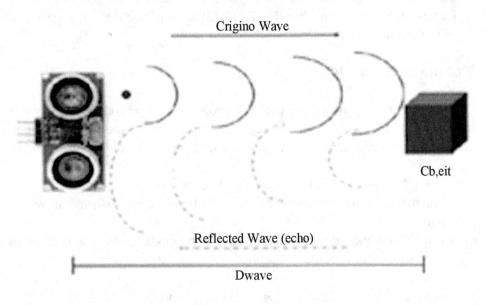

Crigino Wave

Cb,eit

Reflected Wave (echo)

Dwave

Table 2. Comparison table between HC-SR04 and HC-SR05

	HC-SR04	**HC-SR05**
Working Voltage	5 VDC	5 VDC
Static current	< 2mA	<2 mA
Output signal:	Electric frequency signal, high-level 5V, low-level 0V	Electric frequency signal, high-level 5V, low-level 0V
Sensor angle	< 15 degrees	< 15 degrees
Detection distance (claimed)	2cm-450cm	2cm-450cm
Precision	~3 mm	~2 mm
Input trigger signal	10us TTL impulse	10us TTL impulse
Echo signal	output TTL PWL signal	output TTL PWL signal
Pins	1. VCC 2. trig(T) 3. echo(R) 4. GND	1. VCC 2. trig(T) 3. echo(R) 4. OUT 5. GND

Based on the comparison table, SR04 is better in terms of precision for measuring distances, so SR04 is selected for the system measurement.

Figure 12. Temperature and humidity sensor

Module of DHT-11

A temperature and humidity sensor device are shown in Figure 12. DHT11 digital temperature and humidity sensor is the calibrated digital signal output of the combined temperature and humidity sensor. It uses dedicated digital module sensing technology and temperature and humidity sensor technology to ensure products with high reliability and excellent long-term stability. The sensor includes a resistive element and wet NTC device temperature sensors with an attached high-performance 8-bit microcontroller.

HVAC, dehumidifiers, test and control equipment, consumer goods, automotive, automation, data loggers, weather stations, home appliances, humidity controllers, medical and other relevant humidity measurement, and control.

Low cost, long-term stability, relative humidity and temperature measurement, excellent quality, fast response, anti-interference, long-distance signal transmission, digital signal output and accurate calibration. Here the relevant pin of the DHT-11 sensor used for temperature and humidity is shown in Figure 13.

Important parameters of DHT-11 Sensor are Relative Humidity, Resolution 16Bit, Repeatability by $\pm 1\%$RH, accuracy is 25°C with the error of $\pm 5\%$RH, Interchangeability: Fully interchangeable, Response time is1/e (63%) 25°C6s, 1m/

Figure 13. Respective pins of the DHT-11 sensor

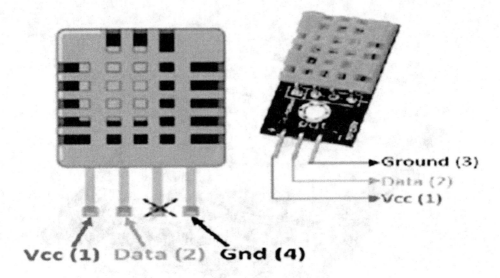

sAir6s and Hysteresis$<\pm0.3\%$R is H along with Long-term stability of $<\pm0.5\%$RH/ yr. Its Temperature Resolution is 16Bit, with Repeatability of $\pm1°$C, Accuracy of $25°$C, and error of $\pm2°$C, along with a Response time of 1/e (63%) 10S.

The selection of DHT-11 for the model is based on its accuracy and low cost. DHT-11 sensor is available in local markets. The effective selection makes the model economical.

Connecting the typical application circuit shown above the microprocessor and DHT11,

DATA pull-up and microprocessor I/O port:-

1. A typical application circuit recommends a cable length shorter than 20 meters with a 5.1K pull-up resistor greater than 20 meters when the pull-up resistor is to reduce the actual situation.
2. When using a 3.3V voltage supply, cable length must not exceed 100cm. Otherwise, it will lead to a lack of line drop sensor supply, causing measurement bias.
3. Each sensor reads the temperature and humidity value in more than 5 seconds to obtain accurate data.

Figure 14. Schematic diagram of DHT-11 sensor

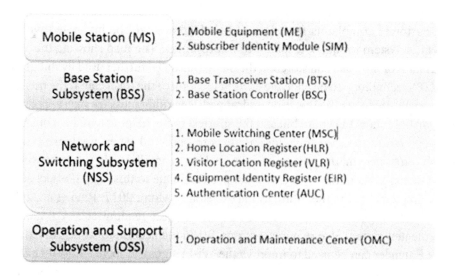

CONCLUSION AND FUTURE ADVANCEMENTS

The chapter presented a practical bin management system for monitoring waste levels. The Smart bin discussed is a real-time waste management system using various sensors to monitor the level and type of waste in the bin. In this system, a GSM module is installed to capture the information about the bin which can be accessed from anywhere at any point of time.

This smart system will help to track and inform the status of each stack in real time. The sensors and IoT allows the waste management team to pick up rubbish when the smart bin is full. The ultra-sonic sensors can help to identify distances between 2 cm and 400 cm. These sensors monitor and compares the depth and intensity of the bin and also measures the level of waste. These sensors will also collect data and send it to the micro-controller for displaying on the LCD Screen.

At the same time, the sensor will share and send data to the web page. The data in the application will display the real-time data. Therefore, the solid waste can be monitored by the management team. However, if the network coverage or internet availability is limited, the Wi-Fi module will be difficult to work. This will disrupt the working of the entire system.

The advancement in development of communication hardware module needs to be improvise in the future. Additionally, the smart system expansion and development can be improved by designing and implementing sensors that can identify different types of trash using image recognition to avoid accidentally putting trash in the wrong bin. Implementation of this proposed system will reduce costs, manpower and indirectly reduce traffic at the site.

A customer complaints module will be added and will be integrated with the SMS alert system for quick response to future works. The map showing the actual coordinates of the trash can stored in the database can be further used by integrating with GPS technology to provide the current location of the trash can. If the trash can has been moved to another location or dropped, a warning message may appear. The map can only show the entire bin and the shortest route, helping to reduce operating costs and emissions. Several limitations were observed when implementing the system. First, the smart trash can should work more intelligently to detect whether the trash can lid is in the closed or open position. Due to this, a Hall effect sensor can be implemented in the future as in (Ramson & Moni, 2017; Ravi et al., 2021), so that the ultrasonic sensor can provide the correct reading at any time. This implemented system also requires a strong Wi-Fi connection to work. An external Wi-Fi Extender can be used to improve the Wi-Fi signal. A local authority can use this type of system to monitor the status of waste collection in real time. Based on the recorded information, they can measure their operational performance, predict future operational requirements and plan for better service delivery.

REFERENCES

Ali, M. L., Alam, M., & Rahaman, M. A. N. R. (2012). RFID-based e-monitoring system for municipal solid waste management. *2012 7th International Conference on Electrical and Computer Engineering*. 10.1109/ICECE.2012.6471590

Amruta, B. P., Madhuri, S., Pallavi, P., & Satish, S. (2018). City Garbage Collection Indicator Using Wireless Communication. *International Journal of Scientific Research in Science, Engineering and Technology*, *4*, 378–380.

Baldo, D., Mecocci, A., Parrino, S., Peruzzi, G., & Pozzebon, A. (2021). A Multi-Layer LoRaWAN Infrastructure for Smart Waste Management. *Sensors (Basel)*, *21*(8), 2600. doi:10.339021082600 PMID:33917255

Bano, A. (2020, December 29). *AIoT-Based Smart Bin for Real-Time Monitoring and Management of Solid Waste*. Https://Www.Hindawi.Com/Journals/Sp/2020/6613263/. https://www.hindawi.com/journals/sp/2020/6613263/

Haque, K. F. (2020, June 1). *An IoT Based Efficient Waste Collection System with Smart Bins*. IEEE Conference Publication. https://ieeexplore.ieee.org/document/9221251

Haque, K. F., Zabin, R., Yelamarthi, K., Yanambaka, P., & Abdelgawad, A. (2020). An IoT-Based Efficient Waste Collection System with Smart-Bins. *2020 IEEE 6th World Forum on Internet of Things (WF-IoT)*. 10.1109/WF-IoT48130.2020.9221251

John, J., Varkey, M. S., Podder, R. S., Sensarma, N., Selvi, M., Santhosh Kumar, S. V. N., & Kannan, A. (2021). Smart Prediction and Monitoring of Waste Disposal System Using IoT and Cloud for IoT Based Smart Cities. *Wireless Personal Communications*, *122*(1), 243–275. doi:10.100711277-021-08897-z

Kumari, P. (2018, January 23). *Iot based smart waste bin model to optimize the waste management process.* http://dl.lib.uom.lk/handle/123/13033

Maheshwaran, K. (2018, April 24). Smart Garbage Monitoring System using IOT. *IJERT*. https://www.ijert.org/smart-garbage-monitoring-system-using-iot

Mamun, M.A., Hannan, M.A., & Hussain, A. (2014). *A novel prototype and simulation model for real time solid wastebin monitoring system.* Academic Press.

Mustafa, M. R. (2017). Smart Bin: Internet-of-Things Garbage Monitoring System. *MATEC Web of Conferences*. Https://Doi.Org/10.1051/Matecconf/201714001030

Praveen, V., Arunprasad, T., & Gomathi, R. N. D. P. (2018). A Survey on Trash Monitoring System using Internet of Things. *International Journal of Trend in Scientific Research and Development, 2*(3), 601–605. doi:10.31142/ijtsrd11020

Ramson, S. J., & Moni, D. J. (2017). Wireless sensor networks based Smart-Bin. *Computers &amp. Electrical Engineering*, *64*, 337–353. doi:10.1016/j.compeleceng.2016.11.030

Ravi, V. R., Hema, M., SreePrashanthini, S., & Sruthi, V. (2021). Smart-Bins for trash monitoring in smart cities using IoT system. *IOP Conference Series. Materials Science and Engineering*, *1055*(1), 012078. doi:10.1088/1757-899X/1055/1/012078

Samann, F. E. F. (2017, August 30). The Design and Implementation of Smart Trash Bin. *Academic Journal of Nawroz University*. Https://Doi.Org/10.25007/Ajnu.V6n3a103. https://journals.nawroz.edu.krd/index.php/ajnu/article/view/103

Sanger, J. B., Sitanayah, L., & Ahmad, I. (2021). A Sensor-based Trash Gas Detection System. *2021 IEEE 11th Annual Computing and Communication Workshop and Conference (CCWC)*. 10.1109/CCWC51732.2021.9376147

Sidhu, N., Pons-Buttazzo, A., Muñoz, A., & Terroso-Saenz, F. (2021). A Collaborative Application for Assisting the Management of Household Plastic Waste through Smart-Bins: A Case of Study in the Philippines. *Sensors (Basel)*, *21*(13), 4534. doi:10.339021134534 PMID:34282802

Srisabarimani, K., Arthi, R., Soundarya, S., Akshaya, A., Mahitha, M., & Sudheer, S. (2022a). Smart Segregation Bins for Cities Using Internet of Things (IoT). *Lecture Notes in Electrical Engineering*, 717–730. doi:10.1007/978-981-16-9488-2_68

Srisabarimani, K., Arthi, R., Soundarya, S., Akshaya, A., Mahitha, M., & Sudheer, S. (2022b). Smart Segregation Bins for Cities Using Internet of Things (IoT). In P. K. Mallick, A. K. Bhoi, A. González-Briones, & P. K. Pattnaik (Eds.), *Electronic Systems and Intelligent Computing. Lecture Notes in Electrical Engineering* (Vol. 860). Springer. doi:10.1007/978-981-16-9488-2_68

Suddul, G. (2018). An Energy Efficient and Low Cost Smart Recycling Bin. *International Journal of Computer Applications*. https://www.ijcaonline.org/archives/volume180/number29/29177-2018916698

Thakker, S., & Narayanamoorthi, R. (2015). Smart and wireless waste management. *2015 International Conference on Innovations in Information, Embedded and Communication Systems (ICIIECS)*. 10.1109/ICIIECS.2015.7193141

Wijaya, A. S. (2017, August 1). *Design a smart waste bin for smart waste management*. IEEE Conference Publication. https://ieeexplore.ieee.org/abstract/document/8068414

Chapter 10

An Unsupervised Traffic Modelling Framework in IoV Using Orchestration of Road Slicing

Divya Lanka
Pondicherry Technological University, India

Selvaradjou Kandasamy
Pondicherry Technological University, India

ABSTRACT

Presenting better traffic management in urban scenarios has endured as a challenge to reduce the fatality rate. The model developed consolidates an evolving methodology to handle traffic on roads in an abstract way in internet of vehicles (IoV) orchestrated with unsupervised machine learning (USL) techniques. At first, the roads are sliced into segments with roadside units (RSU) that provide vehicle to everything (V2X) communication in a multi-hop manner and examine traffic in real-time. The unique nature of the proposed framework is to introduce USL into the RSU to learn about traffic patterns. The RSU upon learning the traffic patterns applies the Gaussian mixture model (GMM) to observe the variation in the traffic pattern to immediately generate warning alerts and collision forward messages to the vehicles on road. The application of USL and GMM ensures speed control of vehicles with traffic alerts, thereby deteriorating the fatality rates.

DOI: 10.4018/978-1-6684-4991-2.ch010

INTRODUCTION

The evolution of new technologies and trends is articulating many innovative applications in diverse fields. Internet of Things (IoT) is one of the most trending concepts that has shown a lot of impact on the lifestyle of human beings and the work nature of industries and organizations. According to the studies 50 million things are involved in the area of IoT. Applying the concept of IoT in the automobile industry led to the development of automotive applications making vehicles smarter. Vehicular Adhoc Networks (VANETS) allow vehicles to communicate with other vehicles treating vehicles as relay nodes. VANETS, IoT, and automotive industries led to the proliferation of Intelligent Transportation Systems (ITS) in smart city applications with the Internet of Vehicles (IoV). IoV empowers vehicles to communicate with other vehicles, infrastructure, road, and pedestrians (vehicle to everything (V2X)) forming an interconnected network called Social Internet of Vehicles (S-IoV). Smart Vehicles in IoV pertain to computing, onboard diagnostics, and communication capabilities. IoV can be applied for a variety of applications like driverless cars, congestion control on roads, monitoring flow of traffic, entertainment, and infotainment services by means of mobile technology, DSRC, Wireless Access for Vehicular Environment, and IEEE 802.11p.

Pedestrians, hurried vehicles on highways, over traveling of trucks and buses were major concerns for fatality rates in India (Mohan et al., 2009). Speed Control and collision alerts are crucial goals to lower the fatality rate. To achieve the goals road segmenting (Abbas et al., 2019) was adapted in the present work to carefully handle the traffic on Indian roads. Roadside units (RSU) are a Dedicated Short-Range Communication (DSRC) device situated aside road ways that acts as entryway between On Board Units (OBU) and Communication infrastructures. RSUs coordinate communication between vehicles (V2V), Vehicle to Cloud (V2C), provide traffic data and sends safety alerts to vehicles.

Traffic Monitoring with unsupervised learning at road segments was proposed in the work. The role of RSU is to manage traffic and various vehicle characteristics. Most commonly Roadside units are restricted to vehicle to vehicle (V2V) communication and vehicle to infrastructure (V2I) communication. High level segmentation of roads is one requirement for the accomplishment of innovative ability. Exhaustive segmentation is relatively rare in both early methodologies. Most recent time's segmentation is mentioned in research articles that insist on significant routing of vehicles to targeted destinations.

In the near future these RSUs can replace the manual process to handle the traffic successfully limiting the congestion on roads and thereby reducing the speed parameters in line with traffic thereby preventing the fatalities caused by accidents. The assumption is that proposed system learns from real time traffic data defining on how vehicles control speed and manage time delays for better traffic management in IoV. The main motivation of the work is to automate the forward collision warnings to the vehicles by RSU's to limit the road accidents. The rest of the work is organized as existing works survey is discussed in section 2, the approach followed in the proposed work is given in section 3 and experimental results are given in section 4.

BACKGROUND

Introducing machine learning in IoV make things to behave like actual things with predictions, clustering and classification. Machine Learning may cut off the limitations in IoV like high mobility, huge volumes of datum and spontaneous decision making (Liang et al., 2019). Machine learning is applied in IoV in the literature to identify misbehavior among vehicles in network, botnets in IoV, controlling traffic in case of emergencies, accidents and vehicle thefts (Dandala et al., 2017). An appropriate machine learning technique for traffic and congestion control in IoV can have good prediction on traffic density on roads with minimum efforts. Ata et al. (2019) presented an automated timing of traffic signals with real time collection of data with artificial neural networks to control traffic congestion with back propagation and time series. Dragon fly optimizer to maintain stable topology by clustering in IoV that vary in traffic density, rapid movement of vehicles is developed by Aadil et al. (2018) to make IoV more scalable. Artificial Intelligence (AI) and Machine Learning (ML) predictions to predict flow of traffic can enhance the efficiency of ITS applications (Boukerche et al., 2020). A new cloud computing through deep learning model is proposed by Chen et al. (2018) to detect collisions in traffic with low response time. To develop the model, Chen et al. (2018) used the sensors in the vehicles periodically to transmit the data to cloud, deep learning methods were run on that data to detect the collisions. The authors Chowdhury et al. (2018), developed a time series-based analysis on past data at road intersections to predict traffic jams on urban roads. Convolution Neural Networks (CNN) is applied to perform image classification on CCTV images to forecast the congestion of traffic. A system was trained with CNN model to foretell jamming at a particular location. Upon receiving a request of traffic condition, CCTV image was forwarded to the trained CNN model to reply on traffic condition (Kurniawan et al., 2018). To minimize the environmental pollution and stress levels of passengers on roads with traffic jams, Long Short-Term Memory (LSTM) cells are used in deep learning on tweets from social media to predict traffic patterns early (Essien et al., 2020). Decision tree (c 4.5) based traffic congestion prediction was also modelled on tweets of traffic and geo tags to plan travel and also tested the model on real time data in Bangkok (Wongcharoen & Senivongse, 2016). The arrival time prediction of public transport vehicles is predicted on twitter information using Kalman filter in (Abidin et al., 2015). A prototype developed with fuzzy logic is proposed by Nguyen et al. (2018) to find traffic congestion with less false predictions between source and destination. For traffic flow prediction a deep belief network was constructed using firefly algorithm with the RSU data for traffic guidance (Goudarzi et al., 2018). Artificial Intelligence, deep learning in ITS can accomplish optimal solutions for enhanced spectrum usage, data security, predictions for improved decision making for the transportation problems with less efforts (Abduljabbar et al., 2019; Sharma et al., 2018; Wang et al., 2020). To track vehicles, to enhance real time communication in IoV, rapid data processing, a hierarchical clustering with satellite terrestrial combined framework is developed in (Lin et al., 2020). USL for identifying intentional and accidental radio frequency signal jamming with clustering effectively works using a novel parameter RF signal speed between receiver and jammer in VANETs (Karagiannis & Argyriou, 2018). From the literature, it was observed that a smaller amount of work was carried on IoV using unsupervised learning.

PROPOSED MODEL

The Three Tier Architecture Model

To model the traffic on roads, three tier architecture was proposed. From Figure 1, we can anticipate that the road is partitioned as segments like Si1, Si2, ... Sin. The vehicles on the road pass through these segments and each segment is connected to corresponding RSU with DSRC between vehicles travelling on roads and RSU's. When the vehicle comes to the vicinity of RSU, the characteristics of vehicles are transmitted to RSU. Timely data at RSUs is transmitted to cloud to perform historical computations as shown in the Figure 1.

S_{i1} to S_{in} =n number of segments in a lane, RSU= Roadside Unit

Network Establishment

The network establishment in IoV is viewed as Bayesian Belief Network as shown in Figure 2. An IoV Bayesian Belief network is built on vehicles as nodes and connectivity to the RSU's and cloud as edges (directed) like Directed Acyclic Graph (DAG). Edge is a directed link representing the relationship between vehicular parameters to connect to the RSU. Each vehicle in IoV Bayesian network holds a conditional probability gives the impact of RSU on that vehicle (Vi). Let IoV includes V1, V2, V3, V4, V5 .... Vn, then the relationship among the vehicles is determined using equation 1.

Figure 1. Three Tier model architecture in IoV

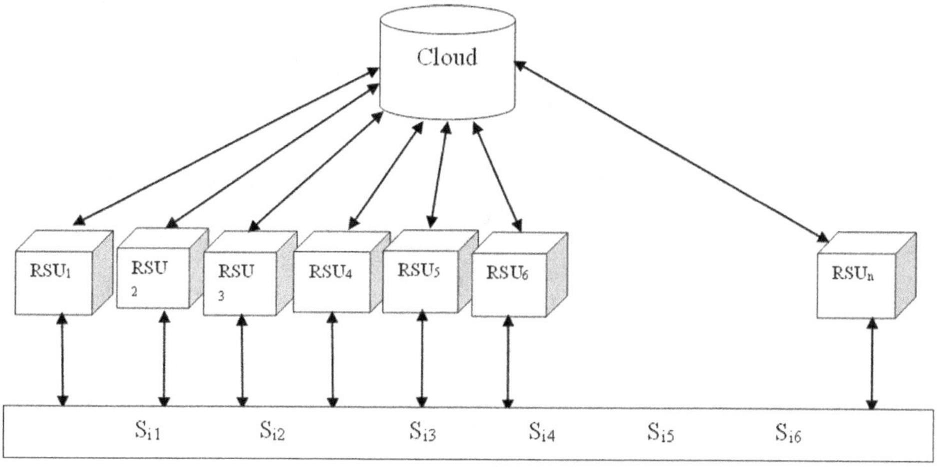

$$P\left(V_{1}, V_{2}, V_{3} \ldots V n\right) = P\left(V_{1} | V_{2}, V_{3} \ldots V n\right) P\left(V_{2}, V_{3}, V_{4} \ldots V n\right) \qquad (1)$$

Road Segmentation Using Unsupervised Learning K-Means Approach

K-means is a tremendous unsupervised learning algorithm. Emphasizing K-means on the IoV is to cluster the vehicles in the corresponding segment, then discover the underlying patterns like speed, traffic flow from vehicular parameters involving vehicle speed, velocity, direction, driving pattern. Road segmentation in IoV refers to the collection of vehicles grouped together relying on point-to-point distance between vehicle to RSU and vehicle to vehicle distance in the related segment. Unsupervised K-Means in IoV considers K roadways named tracks as centroids and upon that every track associate with k segments. The relative distance between the segments is kept small for maintaining good connectivity and supervision of traffic.

RSU is a cooperation of hardware and software solutions to the societies utilizing the IoV networks efficiently on each road segment. In IoV, the major field of innovation is in best traffic management feature which scale the utility of existing RSU's to improve their privacy and functionality by adding features like confidential request handling. The procedure followed is to allocate K segments of roads to K roadways so that the within-segment sum of squares is minimize, randomly the K roadway tracks are chosen at various locations as centroids and upon those random tracks every vehicle joining the network is allocated the closest segment in its path.

Figure 2. IoV Bayesian Belief network formation at RSU with conditional probabilities

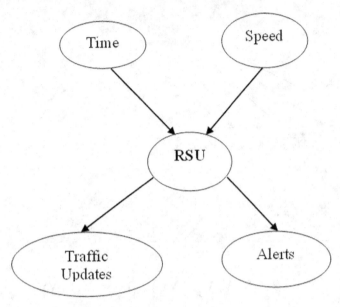

Data learning process in K-Means boots up with random selection of segments. Here in this scenario segment (RSU) is considered as centroid. After random selection of segments for optimized positions of segments the same process is iterated to stabilize the segments. Meanwhile, the number of possible arrangements is enormous; it is not practical to expect the best solution. Rather, this procedure finds a local optimum. This is a solution in which no movement of an observation from one cluster to another will reduce the within cluster sum of squares. The procedure may be repeated several times with different starting configurations. The optimum of these segment solutions is then selected.

IoV comes with heterogeneous and enormous number of vehicles. The K-Means procedure is famous because it can be applied to relatively large sets of data. While deploying the Vehicular network in urban scenarios, the network manager specifies the number of Segments to be found. The procedure then separates the given space into segments by finding a set of segment units, assigning each vehicle to a cluster, determining new segments and repeating this process.

This technique initializes the segments may influences the final segmentation solution. For each request process trail, each vehicle is arbitrarily allocated to a segment. This configuration is optimized using K-Means procedures will greatly increase the probability of finding the global optimum solution for a particular number of clusters.

Goodness-of-fit criterion: The goodness-of-fit criterion shown in Equation 2 is used to compare various cluster configurations based on the sum of squares given as WSSk.

$$WSS_k = \left(\frac{NB}{NB - m} \right) \sum_{k=1}^{k} \sum_{i=1}^{B} \sum_{j=1}^{n_k} \left(1 - \delta_{ijk} \right) \left(z_{ij} - c_{ik} \right)^2 \tag{2}$$

Where c_{ik} is the average center value in the kth cluster.

Generation of Traffic Alerts

Each Segment Monitoring Unit runs over gaussian mixture model with expectation and maximization for efficient traffic management in IoV.

Gaussian Mixture Model: To classify data at each RSU is a challenge. The values at RSU are vehicle parameters like speed, velocity, direction, time while travelling in the corresponding segment are considered. Let X is an arbitrary variable that takes these values. For likelihood model determination we have mixture of Gaussian distribution as the below equations 3 and 4.

$$f(x) = \sum_{i=1}^{c} p_i N\left(x \mid \mu_i, \tilde{A}_i^2 \right) \tag{3}$$

Where c is the number of segments or regions and pi> 0 are weights such that

$$\sum_{i=1}^{c} p_i = 1,$$

$$N(\mu_i, \tilde{A}_i^2) = \frac{1}{\tilde{A}\sqrt{2\tilde{A}}} \, exp\left[\frac{-\left(x - \mu_i \right)^2}{2\tilde{A}_i^2} \right] \tag{4}$$

where $\mu_i, \tilde{A}_i^2$, these two are mean and standard deviation of are class i. For an assumed vehicle, the lattice data are the values of speed, velocity, direction, time in our RSU based model. However, the parameters are $\theta = \left(p_1, \ldots, p_k, \mu_1, \ldots, \mu_k, \sigma_1^2, \ldots, \sigma_k^2 \right)$.

Table 1. Expectation maximization algorithm

Algorithm: Expectation Maximization with E- Step and M- Step
1.Input: Observed vehicle parameters in a vector x_j, j=1,2,...,n and i $\in$ {1,2,...,k}label set.
2. Make ready: $\theta^{(0)} = \left(p_1^{(0)}, \ldots, p_k^{(0)}, \mu_1^{(0)}, \ldots, \mu_k^{(0)}, \sigma_1^{(0)}, \ldots, \sigma_k^{(0)} \right)$
3. E-Step: $$p_{ij}^{(r+1)} = p^{(r+1)}\left(i \mid x_j\right) = \frac{p_i^{(r)} N\left(x_j \mid \mu_i^{(r)}, \sigma_i^{2(r)}\right)}{f\left(x_j\right)}$$
4. M-Step: $$\hat{p}_{ij}^{(r+1)} = \frac{1}{n}\sum_{j=1}^{n} p_{ij}^{(r)}$$ $$\hat{\mu}_i^{(r+1)} = \frac{\sum_{j=1}^{n} p_{ij}^{(r)} x_j}{n\hat{p}_i^{(r+1)}}$$ $$\hat{p}_i^{2(r+1)} = \frac{\sum_{j=1}^{n} p_{ij}^{(r+1)}\left(x_j - \hat{\mu}_i^{(r+1)}\right)^2}{n\hat{p}_i^{(r+1)}}$$
5. Repeat step-2 and step3 until an random error i.e., $\sum_i e_i^2 < \varepsilon$.
6. Calculate the $p_{ij=}Arg\,Max_i\,p_{ij}^{(final)}$ j=1, 2... n.
7. Build traffic modeling corresponding to vehicular characteristics.

Expectation Maximization MapReduce Procedure: Many researchers use the Expectation Maximization (EM) based Map Reduce Procedure for Gaussian Mixture Model. In this regard, this procedure is modified and renamed to Expectation Maximization with MapReduce Procedure. The process of Expectation-Mapping Procedure and Maximization Reduce Procedure steps are shown in Table 1.

EXPERIMENTAL RESULTS

The GMM issues alerts for clustered data with unsupervised machine learning. The difference between K-Means and GMM is that the GMM is a probabilistic model that instead of using distances, it calculates the vehicle speed for classifying data points and uses Gaussian distributions, where each data point is assigned a probability of the likelihood of it being part of a cluster in the model. This is done

Figure 3. Vehicles tested rendering to the time in seconds to run in a particular Segment

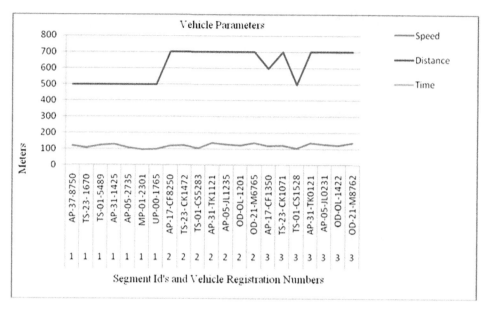

through a single-density estimation method with multiple gaussian probability functions, to generate the probability distribution within the model. For instance, several vehicles were tested according to the time in seconds to run in a particular segment as shown in Figure 3. GMM finds the vehicles with more speed to send warn alerts as shown in Figure 4.

CONCLUSION

In this paper, unsupervised learning-based K-Means and Gaussian Mixture Models are applied on IoV. Firstly, by applying K-Means the vehicles are clustered to segments with RSU's. Vehicles falling within a road segment transmit their extrinsic parameters with RSU's. The RSU runs GMM with EM to select the vehicles to send alert warnings. Receiving prior alert warnings, the vehicles restrict their speed thereby reducing collisions among vehicles and fatality rate on roads. The future enhancements can include identifying of Theft vehicles and social relationship between vehicles. Implementations of machine learning techniques in trusted platform environment at RSU's can prevent generation of fault traffic warning alerts.

Figure 4. Which Vehicle Received Warn Message Re-identifying vehicles

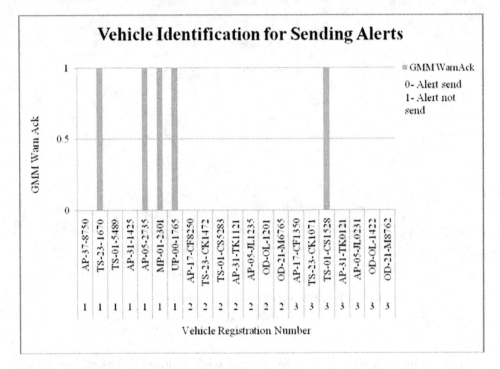

REFERENCES

Aadil, F., Ahsan, W., Rehman, Z. U., Shah, P. A., Rho, S., & Mehmood, I. (2018). Clustering algorithm for internet of vehicles (IoV) based on dragonfly optimizer (CAVDO). *The Journal of Supercomputing*, 74(9), 4542–4567. doi:10.100711227-018-2305-x

Abbas, M. T., Muhammad, A., & Song, W. C. (2019). SD-IoV: SDN enabled routing for internet of vehicles in road-aware approach. *Journal of Ambient Intelligence and Humanized Computing*, 11(3), 1265–1280. doi:10.100712652-019-01319-w

Abduljabbar, R., Dia, H., Liyanage, S., & Bagloee, S. A. (2019). Applications of Artificial Intelligence in Transport: An Overview. *Sustainability*, 11(1), 189. doi:10.3390u11010189

Abidin, A. F., Kolberg, M., & Hussain, A. (2015). Integrating Twitter Traffic Information with Kalman Filter Models for Public Transportation Vehicle Arrival Time Prediction. *Big-Data Analytics and Cloud Computing*, 67–82. doi:10.1007/978-3-319-25313-8_5

Ata, A., Khan, M. A., Abbas, S., Ahmad, G., & Fatima, A. (2019). Modelling smart road traffic congestion control system using machine learning techniques. *Neural Network World*, *29*(2), 99–110. doi:10.14311/NNW.2019.29.008

Boukerche, A., Tao, Y., & Sun, P. (2020). Artificial intelligence-based vehicular traffic flow prediction methods for supporting intelligent transportation systems. *Computer Networks*, *182*, 107484. doi:10.1016/j.comnet.2020.107484

Chen, L. B., Su, K. Y., Mo, Y. C., Chang, W. J., Hu, W. W., Tang, J. J., & Yu, C. T. (2018, September). An Implementation of Deep Learning based IoV System for Traffic Accident Collisions Detection with an Emergency Alert Mechanism. *2018 IEEE 8th International Conference on Consumer Electronics - Berlin (ICCE-Berlin)*. 10.1109/ICCE-Berlin.2018.8576197

Chowdhury, M. M., Hasan, M., Safait, S., Chaki, D., & Uddin, J. (2018, June). A Traffic Congestion Forecasting Model using CMTF and Machine Learning. *2018 Joint 7th International Conference on Informatics, Electronics & Vision (ICIEV) and 2018 2nd International Conference on Imaging, Vision & Pattern Recognition (icIVPR)*. 10.1109/ICIEV.2018.8640985

Dandala, T. T., Krishnamurthy, V., & Alwan, R. (2017, January). Internet of Vehicles (IoV) for traffic management. *2017 International Conference on Computer, Communication and Signal Processing (ICCCSP)*. 10.1109/ICCCSP.2017.7944096

Essien, A., Petrounias, I., Sampaio, P., & Sampaio, S. (2020). A deep-learning model for urban traffic flow prediction with traffic events mined from twitter. *World Wide Web (Bussum)*. Advance online publication. doi:10.100711280-020-00800-3

Goudarzi, S., Kama, M., Anisi, M., Soleymani, S., & Doctor, F. (2018). Self-Organizing Traffic Flow Prediction with an Optimized Deep Belief Network for Internet of Vehicles. *Sensors (Basel)*, *18*(10), 3459. doi:10.339018103459 PMID:30326567

Karagiannis, D., & Argyriou, A. (2018). Jamming attack detection in a pair of RF communicating vehicles using unsupervised machine learning. *Vehicular Communications*, *13*, 56–63. doi:10.1016/j.vehcom.2018.05.001

Kurniawan, J., Syahra, S. G., Dewa, C. K., & Afiahayati. (2018). Traffic Congestion Detection: Learning from CCTV Monitoring Images using Convolutional Neural Network. *Procedia Computer Science*, *144*, 291–297. doi:10.1016/j. procs.2018.10.530

Liang, L., Ye, H., & Li, G. Y. (2019). Toward Intelligent Vehicular Networks: A Machine Learning Framework. *IEEE Internet of Things Journal, 6*(1), 124–135. doi:10.1109/JIOT.2018.2872122

Lin, K., Li, C., Pace, P., & Fortino, G. (2020). Multi-level cluster-based satellite-terrestrial integrated communication in Internet of vehicles. *Computer Communications, 149*, 44–50. doi:10.1016/j.comcom.2019.10.009

Mohan, D., Tsimhoni, O., Sivak, M., & Flannagan, M. J. (2009). *Road safety in India: Challenges and Opportunities.* University of Michigan, Ann Arbor, Transportation Research Institute. https://deepblue.lib.umich.edu/handle/2027.42/61504

Nguyen, D. B., Dow, C. R., & Hwang, S. F. (2018). An Efficient Traffic Congestion Monitoring System on Internet of Vehicles. *Wireless Communications and Mobile Computing, 2018*, 1–17. doi:10.1155/2018/9136813

Sharma, S., Ghanshala, K. K., & Mohan, S. (2018, November). A Security System Using Deep Learning Approach for Internet of Vehicles (IoV). *2018 9th IEEE Annual Ubiquitous Computing, Electronics & Mobile Communication Conference (UEMCON).* doi:10.1109/UEMCON.2018.8796664

Wang, T., Cao, Z., Wang, S., Wang, J., Qi, L., Liu, A., Xie, M., & Li, X. (2020). Privacy-Enhanced Data Collection Based on Deep Learning for Internet of Vehicles. *IEEE Transactions on Industrial Informatics, 16*(10), 6663–6672. doi:10.1109/TII.2019.2962844

Wongcharoen, S., & Senivongse, T. (2016, July). Twitter analysis of road traffic congestion severity estimation. *2016 13th International Joint Conference on Computer Science and Software Engineering (JCSSE).* 10.1109/JCSSE.2016.7748850

Chapter 11
Exploring CNN for Driver Drowsiness Detection Towards Smart Vehicle Development

Pushpa Singh

(iD) https://orcid.org/0000-0001-9796-
3978

*GL Bajaj Institute of Technology and
Management, Greater Noida, India*

Raghav Sharma
KIET Group of Institutions, India

Yash Tomar
KIET Group of Institutions, India

Vivek Kumar
KIET Group of Institutions, India

Narendra Singh

(iD) https://orcid.org/0000-0002-6760-
8550

*GL Bajaj Institute of Technology and
Management, India*

ABSTRACT

*Driver drowsiness is one of the major problems that every country is facing. The
ICT sector is continuously investing in the automaker industry worldwide to bring
about digital transformation in existing vehicles and driving. The smart behavior of
vehicles is becoming possible with the convergence of intelligent manufacturing, AI,
and IoT. In this chapter, the authors are presenting a framework for efficient detection
of driver's drowsiness by utilizing the power of deep learning technology. The use
of convolution neural network (CNN) is explored, and the system is developed and
tested using different activation functions. The proposed driver drowsiness framework
is able to signify the drowsiness state of the driver and to automatically alert the
driver. The accuracy of the proposed model is compared at different activation
functions such as ReLu, SeLu, Sigmoidal, Tanh, and SoftPlus, and higher accuracy
is achieved with ReLu as 98.21%.*

DOI: 10.4018/978-1-6684-4991-2.ch011

INTRODUCTION

Artificial Intelligence, Machine Learning, Deep learning, and IoT revolutionized the healthcare system and gave a new direction to the development of the nation. Disease classification, detection and prediction are increasing the expected life of human beings (Singh et al., 2021). On the other hand, people are losing their lives in road accidents for several reasons. There may be overspeeding, drunk and drive and drowsiness. Drowsiness means feeling unusually sleepy or tired during the day is typically recognized as drowsiness. Drowsiness is a cognitive function of the brain decreases when the person feels sleepy. The condition of drowsiness can be termed as a state of sleep deprivation and tiredness. This drowsiness can hinder the ability to perform any logical or complex tasks of a person. Drowsiness is really problematic and risky when the person is driving. Hence, it is very significant to detect drowsiness. As per our best knowledge, we do not have any concrete practices that convey a direct and clear result of the drowsiness condition.

According to the Central Road Research Institute (CRRI) study, 40% of highway accidents happen just because of drivers' drowsiness 1. According to a survey conducted by National Sleep Foundation, it has been brought in light that about 20% of drivers feel drowsy while driving vehicles (Dua et al., 2021). Real statistics could be much higher, however, as it is difficult to figure out whether a driver was drowsy at the time of the accident. A driver driving in a drowsy state is one of the key factors for road accidents. In "a car safety technology, driver drowsiness detection is essential to prevent road accidents. Nowadays, many people use automobiles for daily commutation, higher living standards, comfortability, and timing constraints to reach destinations. This trend leads to high volumes of traffic in urban areas and highways. In turn, it will raise the number of road accidents with several factors." Primarily, Driver Drowsiness is the main reason for most road accidents. This tally can be lowered by alerting the driver of the drowsy state. Deep learning-based models have the potential to equip vehicles with various driver-assistance technologies.

The deep learning model outperforms in face detection (Hasan et al., 2021) with the ability to automate the feature detection that will be utilized in drowsiness detection through the closeness of eyes. Deep learning algorithms are categorized by the use of neural networks whose models are constructed of huge amounts of layers (Pouyanfar et al.,2018]. There is a specific type of deep neural network (DNNs) called convolutional neural networks (CNNs), which have great performance on computer vision due to its ability to detect the pattern and identify characteristics among images (Magan et al. 2022 and Luo et al., 2018). CNN is a kind of deep neural network mostly used to analyze visual imagery (Li et al., 2021). CNN is based on a particular type of method known as Convolution. Convolution is a mathematical process based on two functions that result from a third function which describes that the shape of one is changed by the other. The role of CNN is to minimize the image shape that can be easily processed by any digital device without losing the principal component required for getting a good prediction. Activation function determines what is to be fired or transferred to the next layer of neuron. Various activation function affects the performance of neural network model. Hence, it is important to identify which activation function is enhancing your system accuracy (Ertuğrul, 2018).

In this chapter, the discussion regarding the avoidance of accidents due to drowsiness is discussed with the help of the measure of the eye closure and a corresponding system is developed using CNN. Currently, transport systems play a crucial role in the human economy. This chapter aims to suggest the best possible scenarios to increase or to find a better and more efficient way to detect the fatigue level of the driver. By analyzing the state of the human eye, it can be concluded that the signs of driver fatigue are present early enough to prevent a possible road accident, which could result in injuries or, ultimately, in fatalities.

The remaining chapter is prepared as follows. Section 2, surveys the related work to identify the accuracy of a different model. Section 3 is the methodology that highlights the proposed overall work. The result is represented in section 4, and lastly, section 5 concludes the chapter.

BACKGROUND

Various texts have been available in order to identify driver drowsiness. Mardi et al. (2011) suggested a model to identify drowsiness based on Electroencephalography (EEG) signals. The author recorded EEG signals of 10 volunteers and evaluated them by ANN with 83.3% accuracy. Dwivedi et al,. [2014] projected a CNN model to detect the driver drowsiness based on only visual facial features proficiently on various datasets with average accuracy 78%. Noori et al., (2014) suggested a model detect drowsiness based on the fusion of Driving Quality Signals, EEG and Electrooculography. A self-organized map of the network was applied for the classification and attained an accuracy of 76.51 ± 3.43 per cent.

Park et al. (2016) applied three deep neural networks to achieve universal robustness to related environmental variations and facial detection methods significant for a reliable state detection. The results of the three model is fed into Softmax, and the accuracy of their model was 73.06% on the NTHU-drowsy driver detection benchmark dataset. Zhang et al., projected a system based on the detection of yawn, blink and BVP for the identification of drowsiness. Their work was motion-tolerant to typical head motions; however, the system suffered from the drastic change of head motion.

Chirra (2019) provided deep CNN techniques based on Eye state while driving the vehicle. The viola-Jones face detection algorithm is used to identify the face area and extract the eye region. A SoftMax layer in the CNN classifier was used to categorize the drowsiness case of the driver with 96.42% accuracy. Dua et al., suggested architecture for drowsiness detection with four deep learning models with 85% accuracy.

Arefnezhad et al. (2022) suggested a framework based on EEG signal with dynamical encoder-decoder modelling for drowsiness detection. RMSE and HPD percentage metrics was used to estimate the performance of the projected encoding–decoding model. The average RMSE was 0.117 and average HPD was 62.5%. Further, Magan et al., 2022 applied two different mechanisms to minimize the false-positive rate. The first mechanism applied recurrent and CNN, and the second mechanism was deep learning method to retrieve numeric features from the provided images. Both the systems obtain the accuracy is almost the same: 65% accuracy for training data and 60% accuracy for the test data. Various methods are available to detect faces, as represented in table 1.

We require a technique or model that can identify faces with higher accuracy. This motivates authors to propose a model to enhance accuracy.

MAIN FOCUS OF THE CHAPTER

Drowsiness detection is an important aspect to reduce the chances of accidents. Devices enabled with high pixel camera capture the image of driver face. This camera is fixed in front of the driver seat. After the face detection system apply face detection method. There is the various method available to detect faces. We require a technique or model that can identify faces with higher accuracy. The foremost idea of drowsiness detection is to capture the eyes of the driver from a camera and be able to precisely compute the score of that mark as an indicator for drowsiness. The overall methodology is shown in figure 1.

The proposed model aims to provide a method to detect drowsiness of drivers with higher accuracy. Model is trained with Deep learning model CNN and data set obtained from Kaggle. Model is compared with different activation functions such as Sigmoidal, SeLu, ReLu, SoftMax, etc. and compared the accuracy.

Software and Libraries Used

Free and open-source software is used to design this system. Following are some important software tools are used in designing the proposed model.

Table 1. Important Findings of Related Work

Work	Contribution	Method	Accuracy
Mardi et al., (2011)	Find a drowsiness based on EEG	ANN	83.3%
Dwivedi et al. (2014)	Applied a 2D CNN on each frame for feature extraction and drowsiness detection.	CNN	78%
Noori et al., (2016)	Combination of driving quality signals, EEG and Electrooculography	Self-organized map network	76.51±3.43%
Park et al. (2016)	Three networks were combined and result are fed into a softmax classifier for drowsiness detection	Deep network	73.06%
Zhang et al. (2019)	Identification of yawn, blink and BVP for the identification of drowsiness.	-	-
Chirra et al, (2019)	Viola-Jones face detection algorithm	deep CNN with SoftMax	96.42%
Kavitha et al., (2021)	Naïve Bayes as Face detection and ANN for Drowsiness detection	ANN	N/A
Dua et al., (2021)	Implement four deep learning model	AlexNet, VGG-FaceNet, FlowImageNet and ResNet	85%
Arefnezhad **et al. (2022)**	Drowsiness detection model based on EEG signal with a dynamical encoder–decoder	Encoding–decoding	62.5%
Magan et al., (2022)	Minimization of false positive rate	RNN and CNN	60%

OpenCV: It is a free and open-source library. It offers image processing and computer vision task. In the proposed work, we utilized its function for face detection.

Keras: Keras provides an interface for Python artificial neural networks and the TensorFlow library. It offers the functions of an artificial neural networks. Keras layers are the basic building blocks of neural networks in it.

Numpy: NumPy provides large, multi-dimensional matrices and arrays and a vast collection of mathematical functions.

Dataset Description

The proposed research will be worked on the "Drowsiness detection" dataset. The dataset has been acquired from Kaggle (https://www.kaggle.com/code/chaimaemoumou/drowsiness-detection). The data set consists of a total of 1452 images which consists of different ethnicity of people's eye images under different lighting conditions. The dataset consists of two categories, ' Open' and 'Closed'. The model has been trained by a combined training set of 600 Open Eye Images with

600 Closed Eye Images. The data was manually prepared by deleting and filtering the unnecessary images.

Figure 1. Proposed Methodology

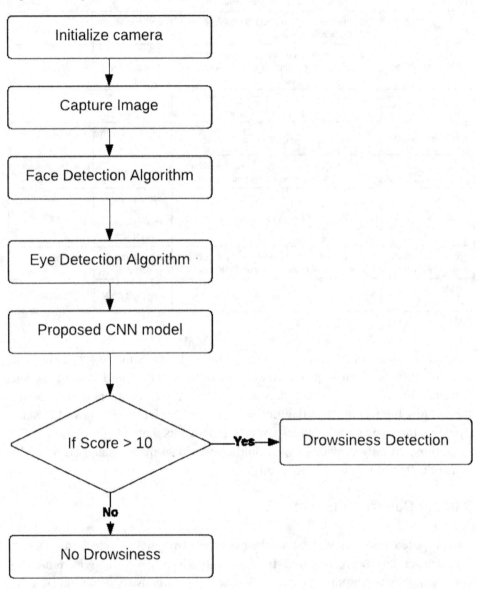

Proposed CNN Model

CNN has great potential to design the projected model to recognize driver drowsiness. A feature vector is required to compare the drowsy image with the face of the driver to detect whether the driver is in a drowsy condition. Usually, CNN models require fixed-sized images as input so pre-processing of images is essential. The pre-processing comprises pulling out the keyframes from the video and store in the database. These feature vectors are then used the detect the driver's drowsiness.

The proposed CNN model is represented in figure 1. A total of 1200 images are used for training the model, and a total of 252 images are used for the validation of the model. A different layer of the CNN model is shown in figure 2.

CNN model is based on the following layers:

1. Convolutional layers
2. Pooling layers
3. Activation function layer
4. Fully connected layer.

The convolution layer has filters, and each kernel has height, depth and width, and this layer produces the feature maps. Pooling layers (Max or Average) of CNN are used to reduce the feature maps' size to speed up computations. The input image is partitioned into various parts then operations are applied on each part of this layer. In Max Pooling, a maximum value is selected for each region of a layer and placed it in the equivalent place of the output.

We are using 5 layers of activation. The first four layers are activated by the following functions: Relu, Tanh, Selu, Softplus, and Sigmoid. The 5th layer is being activated by the activation function called 'softmax' to squash the matrix into output probabilities. Further, we train the model and compile it with the help of the 'Adam' optimizer.

Face and Eye Detection Algorithm

For the detection of drowsiness, one has to capture the eyes and from the face. Various face detection algorithms are available such as Haar cascade, HOG + Linear SVM, SSD, YOLO, etc. OpenCV's Haar cascade method is applied to detect the face for the proposed work. The eye-detection algorithm extracts the eye part from the images and is given as input to CNN when the face is seen. The softmax layer in CNN categorizes the images into drowsy or non-drowsy images.

Figure 2. Proposed CNN Model

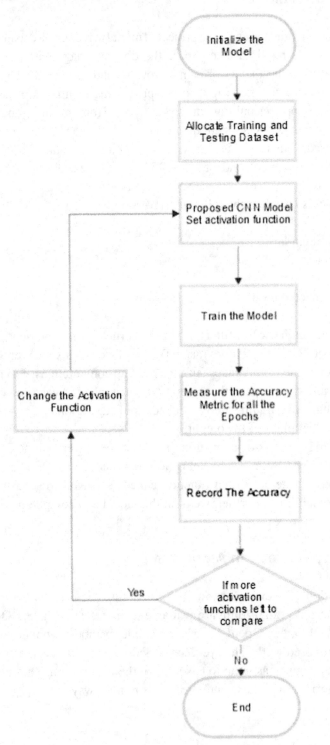

Figure 3. Layers of CNN Model

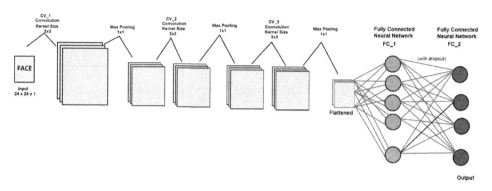

Activation Function

The activation function layer in the neural network describes the way of the weighted sum of the input is changed into an output from a node or nodes in a layer of the network. Activations functions are used for the comparison of the accuracy of the model. Some important activation functions are ReLu, Sigmoid, SeLu, Tanh, Softplus, etc. Dubey et al., in 2021, offers a study of these activation function. Following Activations functions are applied for the comparison accuracy of the proposed work and also represented in table 2.

Relu: The rectified linear activation function is a piecewise linear function that will output the input straight if it is positive, otherwise, it has output zero.

Sigmoid: The sigmoid activation function is also known as logistic function. Sigmoidal activation function is traditionally a well-known activation function for neural networks. The input to the function is altered into a value between 0.0 and 1.0.

Selu: Scaled Exponential Linear Units, or SELUs, are activation functions that induce self-normalizing properties. The SELU activation function is given by. f (x) = λ x if x $\geq$ 0 f (x) = λ α (exp $f(x)$ with $\alpha \approx$ 1.6733 and $\lambda \approx$ 1.0507.

Tanh: Tanh is hyperbolic tangent activation function and it is like to the sigmoid activation function. The function takes any real value as input and outputs values in the range -1 to 1.

Softplus: The softplus function is a smooth approximation to the ReLU activation function. Softplus function is occasionally used in the neural networks instead of ReLU function. softplus(x)=log(1+ex) It is closely related to the sigmoid function. As x$\to-\infty$, the two functions become identical.

Table 2. Important Activation Function (Singh et al.,2021)

Activation Function	Formula	Shape
Linear	$f(x) = cx$	
Sigmoidal	$f(x) = \dfrac{1}{1+e^{-x}}$	
Tangent	$f(x) = \dfrac{e^x - e^{-x}}{e^x + e^{-x}}$	
ReLu	$f(x) = \max(0,x)$	
Softmax	$f_i(x) = \dfrac{e^{x_i}}{\sum\limits_{k} e^{x_k}}$	

Figure 4. No Drowsiness

Drowsiness Detection

As it can be clearly seen in Figure 4, the model detects the face of a person. The status and the score of the person can be seen in the bottom left corner of figure 4 and figure 5, the eyes are closed, and the status has been changed to close from open. These figures are captured by the prepared model during testing. The first figure shows the person being captured by the camera and it's face being detected by the camera. Further the eyes are constantly monitored to record the eye closure. In the figure 4(b) we can see the eyes being closed. The red border indicates the alarm being turned on as the score has crossed the set mark of 10 in figure 4(b). The score is being calculated by measuring each frame the eyes are closed.

score = No. of frames eyes are closed

RESULTS AND DISCUSSION

The datasets have been used to design a model and make comparisons on how different activation functions affect the accuracy of the model, i.e., the detection of the facial features. The results we have produced at 11 epochs. The results greater than 11 epochs led the model towards overfitting and resulted in decreased accuracy. In order to get more accurate results, we needed to deliver and run more epochs. We have used different activation functions to study the accuracy change upon changing the neurons' activation function in the neural network. Table 3 shows the accuracy results of the model measured with the corresponding functions being used to train the model.

Figure 5. Drowsiness Detection

Table 3. Accuracy result with different activation functions

S. No.	Activation Function	Epoch No.	Accuracy
1.	Relu	5	98.21%
2.	Selu	10	93.30%
3.	Tanh	3	91.96%
4.	Softplus	1	52.23%
5.	Sigmoid	5	51.34%

Result is presented in figure 6. Result demonstrates that Relu activation gives the best result when it comes to face detection on this dataset. From the above result we can conclude that the Relu activation function combined with the Softmax function at the end to squash the matrix into output probabilities gives the maximum accuracy.

Limitation of proposed model depend upon accuracy of algorithm and timely provide alert information to the driver/user. Since, right information at right time can make the model efficient and accurate. Major development in this view consists 6G technologies which has potential to enhance user experience in term of high data rate and low latency. Next generation network such as 6G with emerging technologies have potential to provide to high data transmission rate services, ultra-low latency, always best connected and secure network (Singh et al., 2022; Verma et al., 2021).

Figure 6. Comparison of Accuracy

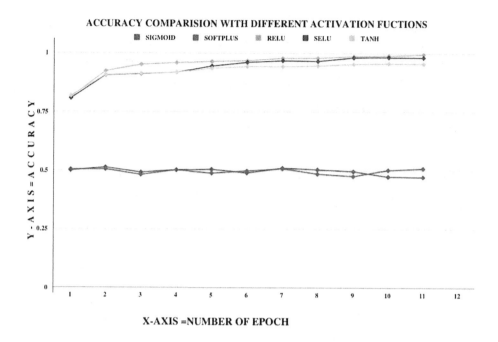

X-AXIS =NUMBER OF EPOCH

CONCLUSION

Our study on the comparison of the various activation functions in the driver drowsiness detection system enables us to improve the accuracy of the model. The various combinations of the activation functions gives more room of improvement to the currently established driver drowsiness detection models. The eyes are seized by utilizing a haar cascade classifier. The design we have built with the help of Convolutional Neural Networks (CNN). The relu activation function along with the adam optimizer yields the maximum accuracy, and hence our model detects the drowsiness condition of the driver more accurately and efficiently than if it had been any other activation function like tanh, softplus, sigmoid or selu. The other functions cannot activate the neurons of the CNN layers as effectively as relu does for this dataset. This provides the future scope of the current and upcoming models of Driver Drowsiness detection systems to be more precise and accurate.

REFERENCES

Arefnezhad, S., Hamet, J., Eichberger, A., Frühwirth, M., Ischebeck, A., Koglbauer, I. V., Moser, M., & Yousefi, A. (2022). Driver drowsiness estimation using EEG signals with a dynamical encoder–decoder modeling framework. *Scientific Reports*, *12*(1), 1–18. doi:10.103841598-022-05810-x PMID:35173189

Chirra, V. R. R., ReddyUyyala, S., & Kolli, V. K. K. (2019). Deep CNN: A Machine Learning Approach for Driver Drowsiness Detection Based on Eye State. *Rev. d'Intelligence Artif., 33*(6), 461-466.

Dua, M., Singla, R., Raj, S., & Jangra, A. (2021). Deep CNN models-based ensemble approach to driver drowsiness detection. *Neural Computing & Applications*, *33*(8), 3155–3168. doi:10.100700521-020-05209-7

Dubey, S. R., Singh, S. K., & Chaudhuri, B. B. (2021). *A Comprehensive Survey and Performance Analysis of Activation Functions in Deep Learning*. arXiv preprint arXiv:2109.14545.

Ertuğrul, Ö. F. (2018). A novel type of activation function in artificial neural networks: Trained activation function. *Neural Networks*, *99*, 148–157. doi:10.1016/j. neunet.2018.01.007 PMID:29427841

Hasan, M. K., Ahsan, M., Newaz, S. H., & Lee, G. M. (2021). Human face detection techniques: A comprehensive review and future research directions. *Electronics (Basel)*, *10*(19), 2354. doi:10.3390/electronics10192354

Kamran, M. A., Mannan, M. M. N., & Jeong, M. Y. (2019). Drowsiness, fatigue and poor sleep's causes and detection: A comprehensive study. *IEEE Access: Practical Innovations, Open Solutions*, *7*, 167172–167186. doi:10.1109/ACCESS.2019.2951028

Kavitha, M. N., Saranya, S. S., Adithyan, K. D., Soundharapandi, R., & Vignesh, A. S. (2021, November). A novel approach for driver drowsiness detection using deep learning. In. AIP Conference Proceedings: Vol. 2387. *No. 1* (p. 140027). AIP Publishing LLC. doi:10.1063/5.0068784

Li, Z., Liu, F., Yang, W., Peng, S., & Zhou, J. (2021). A survey of convolutional neural networks: Analysis, applications, and prospects. *IEEE Transactions on Neural Networks and Learning Systems*, 1–21. doi:10.1109/TNNLS.2021.3084827 PMID:34111009

Luo, H., Xiong, C., Fang, W., Love, P. E., Zhang, B., & Ouyang, X. (2018). Convolutional neural networks: Computer vision-based workforce activity assessment in construction. *Automation in Construction, 94*, 282–289. doi:10.1016/j.autcon.2018.06.007

Magán, E., Sesmero, M. P., Alonso-Weber, J. M., & Sanchis, A. (2022). Driver Drowsiness Detection by Applying Deep Learning Techniques to Sequences of Images. *Applied Sciences (Basel, Switzerland), 12*(3), 1145. doi:10.3390/app12031145

Mardi, Z., Ashtiani, S. N., & Mikaili, M. (2011). EEG-based somnolence detection for safe driving exploitation chaotic options and applied math tests. *Journal of Medical Signals and Sensors, 1*(2), 130–137. doi:10.4103/2228-7477.95297 PMID:22606668

Noori, S.M., & Mikaeili, M. (2016). Driving temporary status detection victimization blending of electroencephalography, electrooculography, and driving nature signals. *J Master's Degree Signals Sens, 6*, 39-46. doi:10.4103/2228-7477.175868

Park, S., Pan, F., Kang, S., & Yoo, C. D. (2016, November). Driver drowsiness detection system based on feature representation learning using various deep networks. In *Asian Conference on Computer Vision* (pp. 154-164). Springer.

Pouyanfar, S., Sadiq, S., Yan, Y., Tian, H., Tao, Y., Reyes, M. P., Shyu, M. L., Chen, S. C., & Iyengar, S. S. (2018). A Survey on Deep Learning: Algorithms, Techniques, and Applications. *ACM Computing Surveys, 51*, 1–36.

Singh, P., Singh, N., Singh, K. K., & Singh, A. (2021). Diagnosing of disease using machine learning. In *Machine Learning and the Internet of Medical Things in Healthcare* (pp. 89–111). Academic Press. doi:10.1016/B978-0-12-821229-5.00003-3

Singh, P., Singh, N., & Deka, G. C. (2021). Prospects of machine learning with blockchain in healthcare and agriculture. In *Multidisciplinary functions of Blockchain technology in AI and IoT applications* (pp. 178–208). IGI Global. doi:10.4018/978-1-7998-5876-8.ch009

Singh, P., Singh, N., Ramalakshmi, P., & Saxena, A. (2022). Artificial Intelligence for Smart Data Storage in Cloud-Based IoT. In *Transforming Management with AI, Big-Data, and IoT* (pp. 1–15). Springer. doi:10.1007/978-3-030-86749-2_1

Singh, M. K., Singh, R., Singh, N., & Yadav, C. S. (2022). Technologies Assisting the Paradigm Shift from 5G to 6G. In *AI and Blockchain Technology in 6G Wireless Network* (pp. 1–24). Springer. doi:10.1007/978-981-19-2868-0_1

Verma, A., Singh, P., & Singh, N. (2021). Study of blockchain-based 6G wireless network integration and consensus mechanism. *International Journal of Wireless and Mobile Computing*, *21*(3), 255–264. doi:10.1504/IJWMC.2021.120906

Zhang, C., Wu, X., Zheng, X., & Yu, S. (2019). Driver drowsiness detection using multi-channel second order blind identifications. *IEEE Access: Practical Innovations, Open Solutions*, *7*, 11829–11843. doi:10.1109/ACCESS.2019.2891971

Chapter 12

Workplace Automation for Total Security Using Voice Activation and Smart Touch

Sachin Kumar Yadav
HMRITM, Delhi, India

Devendra Kumar Misra
ABES Engineering College, India

ABSTRACT

A new era of technology has evolved over the years. Reliability has grown because of its accessibility, speed, and efficiency. This has enabled us to maintain the environment ranging from reducing pollution to low consumption of electricity. Nowadays we are relying so much on digital gadgets that it has started to play an important role in our daily lives, as these digital gadgets can connect with many other devices, making us perform our tasks more easily. With the help of the home automation system, we can not only connect several different devices, but we can also reduce the power consumption. The hallmark of home automation in this proposed work is a remote control, which is done either through a mobile application or a voice assistant. In this system, the mobile application is used for the Android system and for connectivity, Bluetooth has been used as the medium to control the devices. The mobile application allows them to control the devices in real-time. In addition, a capacitive switch that replaces the toggle switch has been used.

DOI: 10.4018/978-1-6684-4991-2.ch012

INTRODUCTION

The primary goal of a home automation system is to automatically or remotely control one or more basic home appliances. It is designed to control and access devices like smart locks, security cameras, smart refrigerators, washing machines and many more. Home automation has made day to day tasks easier and faster. Automation and wireless technology have reduced the need for cables, which makes the system more secure and makes a mark not only in the digital world but also in people's lives (Pandya et al., 2016);(Raheem et al., 2017);(Kannapiran et al., 2017). The interesting part of home automation systems is that the devices can be controlled via a smartphone, making the system more efficient and accessible at one's fingertips. Connecting devices via smartphone has become more convenient for older adults and physically challenged people as it can help them to access and control devices independently while sitting in one place without the assistance of another person. Also, people can change the setting of their system according to their need. Other wireless technologies used in home automation are the Internet, Cloud, GSM and Bluetooth. In the chapter, Bluetooth has been used as a medium to control the devices. Although each technology has its advantages and disadvantages, there is a way to overcome the disadvantages in a Bluetooth based home automation system(Ishak et al., 2014). The typical range of Bluetooth to connect is between 10m and 100m. Nevertheless, it can be increased using the Pico-Nate architecture, and the frequency used for Bluetooth is 2.4GHz, which is available globally. So the speed that can be achieved for Bluetooth services is up to 3 Mbps, and this fundamental advantage was what prompted us to work on Bluetooth-based home automation(Patel et al., 2021);(Amoran et al., 2021);(Budiharto et al., 2019).

Home automation systems are being used and implemented more frequently because they enhance residents' quality of life by bringing them comfort, convenience, and security. Nowadays, the major concern of home automation systems is to reduce the human labor in the production of services and goods hence provide ease to disabled and elderly people. A home automation system can be designed and developed using a single controller that can manage and operate a variety of linked products. Power plugs, lighting, temperature and humidity sensors, smoke, gas, and fire detectors, as well as security systems, can all be used with the system(Raheem el al., 2017). The ability to effortlessly operate, monitor, and manage a home automation system from a variety of devices, including a smartphone, tablet, desktop, and laptop, is one of its major benefits (Kannapiran et al., 2017). In figure 1 a smart home is showing. The wireless communication techniques used by smartphones to communicate with microcontrollers include GSM (Hisham, 2014), Bluetooth (Patel et al., 2021), ZigBee(Amoran et al., 202) and Wi-Fi. Home automation systems offer a variety of services and applications, including appliance control and monitoring, thermostat

Figure 1. Smart Home
(Budiharto et al., 2019)

control, live video surveillance, security camera monitoring, real-time text alerts, and so on. With Android application technology, a remote controlled home automation system provides simple resolution. Any smartphone/tablet, etc., running Android OS can perform remote control via a graphical user interface-based touch screen operation.

This chapter can be operated in three ways- Voice Activation, Android Smartphone System and Smart Touch Switch for home automation applications.

OVERVIEW OF THE SMARTHOME

Figure 2 displays the fundamental block diagram for the home automation system. The values of physical conditions are obtained via sensors that are wired to the micro-controller (Budiharto et al., 2019); (Ransing et al., 2015);(Felix et al., 2011). These integrated sensors can detect smoke and cooking gas to prevent fires from starting, and the temperature sensor is technology-oriented, low-cost, and simple to scale. The Light Dependent Resistor (LDR), which determines the day light intensity, controls the light's on/off switch. A motion detector is integrated to incorporate security in the design using Passive Infrared Sensor (PIR) to turn the security system.

Figure 2. Block Diagram of the Home Automation System
(Felix et al., 2011)

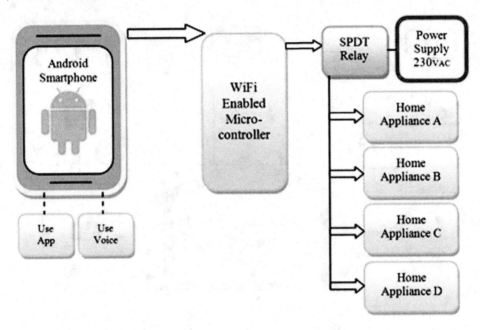

The micro-controller uses a relay switch to transmit control signals to the electronic component that performs the turning on and off action. An authentication system is built into a web portal to verify the user's authenticity (username and password).

TECHNOLOGIES AND SYSTEMS

1. Bluetooth Based Home Automation System

Home automation systems using Bluetooth, Arduino boards, and smartphone applications are affordable and safe. The hardware architecture of the home automation system proposed in (Ramlee et al., 2013) consists of an Arduino BT board and a smartphone. The Arduino BT board and the phone can communicate wirelessly because to the usage of Bluetooth technology. The range, data rate, and bandwidth of the Arduino BT board are as follows: 10 to 100 metres, 3 Mbps, and 2.4 GHz. Relay issued to connect home appliances with the Arduino BT board. The android application used in cell phone which the user to control the home appliances. For the security purposes password based protection is implemented to check the

Figure 3. Block diagram for Bluetooth based HAS
(Chinchansure et al., 2014)

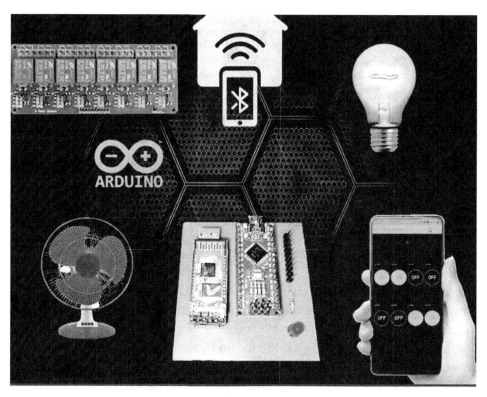

authenticity of the user. The block diagram of a Bluetooth-based home automation system (HAS) is shown in Figure 3.

This home automation system is compatible with automated systems and existing residences. The Only linked appliances in the Bluetooth range can be controlled by the system.

A smart living system that is affordable, user-friendly, and managed by android applications is shown by similar study by (Ullah et al., 2016). Android devices and household appliances can wirelessly connect because of Bluetooth-based technologies for the proposed smart living systems. It also places a strong emphasis on security and alert systems.

2. GSM Based Home Automation System

GSM (Global System for Mobile Communication) deployment results in the creation of a smart home automation system (Chinchansure et al., 2014); (Ullah et al., 2016). The study project places special emphasis on the possibility for complete house

Figure 4. Block diagram for GSM based HAS
(Ullah et al., 2016)

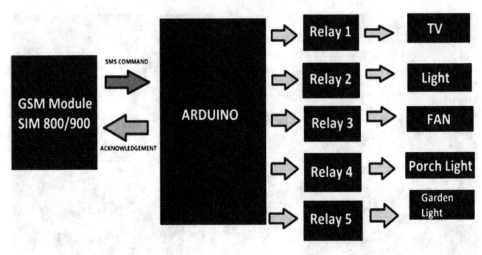

control, which will soon be the main objective of the home automation system. Short Message Service (SMS) text messages are used to operate home appliances including lighting, climate control, and security systems as part of the home automation system, which is built utilizing GSM modem technology. The suggested research project discussed in the paper focuses on the GSM protocol's capabilities, which enables the user to remotely operate the target system via frequency bandwidths. The user can obtain the status of any home equipment through a feedback system that has also been built.

The system's GSM modem, PIC16F887 microcontroller, and smartphone make up its hardware architecture. Relays are used to link home appliances with the PIC16F887 microcontroller. The GSM modem and microcontroller communicate serially using the RS232 protocol. The entire operation (sending and receiving commands) is completed by the GSM modem in less than 2 seconds thanks to its reaction time of less than 500 microseconds. The block diagram of a GSM-based home automation system (HAS) is shown in Figure 4.

Advantage: 1) The user will receive feedback status of household appliances on their smartphones via SMS. 2) The accuracy achieved is greater than 98%. 3) Due to the extensive GSM coverage, customers can access appliances from anywhere in the world. 4) GSM technology in a home automation system offers the highest level of security and dependability.

Similar research carried out in (Ullah et al., 2016), employs a Smartphone application, an LCD, a microprocessor LPC2148, and a GSM SIM900 module as its user interface. The user sends a message from an android application to GSM

Figure 5. Block diagram for voice recognition HAS
(Ullah et al., 2016)

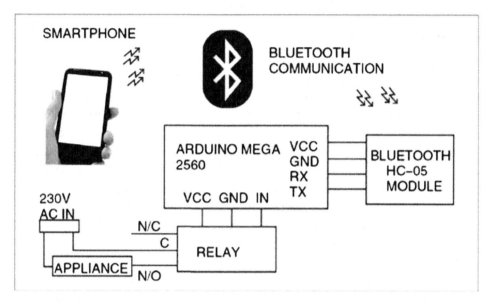

SIM900 module for controlling the home appliances. A add on feature to this system is an LCD display which displays the important notifications.

3. Voice Recognition-Based Home-Automated System

In (Ullah et al., 2016) consists of Arduino UNO and a smartphone. It is a voice recognition-based system, which uses Bluetooth technology for wireless communication between the components. A smartphone application that enables voice control of home appliances was created using the Android OS's built-in voice recognition function. The user's voice command is converted into text by this application, which then sends the text message to the Bluetooth module HC-05 attached to the Arduino UNO. Figure5 illustrates the block diagram of voice recognition-based HAS.

Advantage: It reduces the human effort largely that the user is required to only speak the appliance's name into the microphone to turn the device ON or OFF.

Limitation: 1) The use and efficiency of the system is limited to the range of Bluetooth, though the use of internet can extend the range of the system but that will not be cost effective. 2) The system fails to work efficiently in a noisy environment.

Another research done by(Ullah et al., 2016) use GPRS technology for home automation system using voice recognition. The user of this system can utilize voice commands to operate household equipment. Support Vector Machine (SVM), a learning classifier, is employed in this system machine to recognize voice.

4. ZigBee Based Wireless HAS

The Wireless Sensor Network (WSN) is an independent spatial sensor network used to sense certain tasks. Similar to the ZigBee network, WSN enables the interconnection of straightforward, low-power, and low-processing capable wireless devices. Numerous applications, including pervasive computing, security monitoring, and control, are made possible by ZigBee devices. Sensing data is gathered by ZigBee end devices and transmitted to the coordinator. The coordinator responds to requests from end devices. The performance of the network as a whole may be negatively impacted by a high number of random non-synchronized requests. To create a dependable ZigBee network, an efficient method is especially required for synchronizing available node request processing.

A study in (Ullah et al., 2016) emphasizes on ZigBee-based home automation system with an emphasis on new approaches to planning and simulating a vast ZigBee network. The handheld microphone module, central controller module, and appliance controller module are the three primary modules that make up the ZigBee-based home automation system that is the subject of this study. The central controller module is based on a PC, while the handheld microphone module uses the ZigBee protocol. For the purpose of voice recognition, Microsoft speech API is used and because of low power consumption and cost efficiency, RF ZigBee modules are used for wireless network establishment. A voice of sampling frequency 8 KHz is recorded by the system. The frequency range used for the encoding is 6 Hz to 3.5 KHz. The ZigBee communication protocol offers a maximum baud rate of 250 Kbps, while the microcontroller uses 115.2 Kbps for sending and receiving data. The microcontroller delivered these data bits to the RF ZigBee module at the maximum rate of 115200 bits/s. This automation system is tested using 35 male and female voice commands with various English accents. A total of 1225 speech commands were tested, and the system accurately identified 79.8% of them. Each user recorded 35 voice samples for this purpose. The system's accuracy is impacted by the speaker's accent, speech rate, and ambient noise. The block diagram of a ZigBee-based home automation system (Huang et al., 2010) ;(Piyare et al., 2011);(Kumar et al., 2014); (Teymourzadeh et al., 2013) is shown in Figure 6.

Limitation: This system's accuracy is restricted to a 40-meter range, but a recognition system with a clean line of sight transmission can be accurate up to 80 meters.

Figure 6. Block diagram for voice ZigBee based HAS

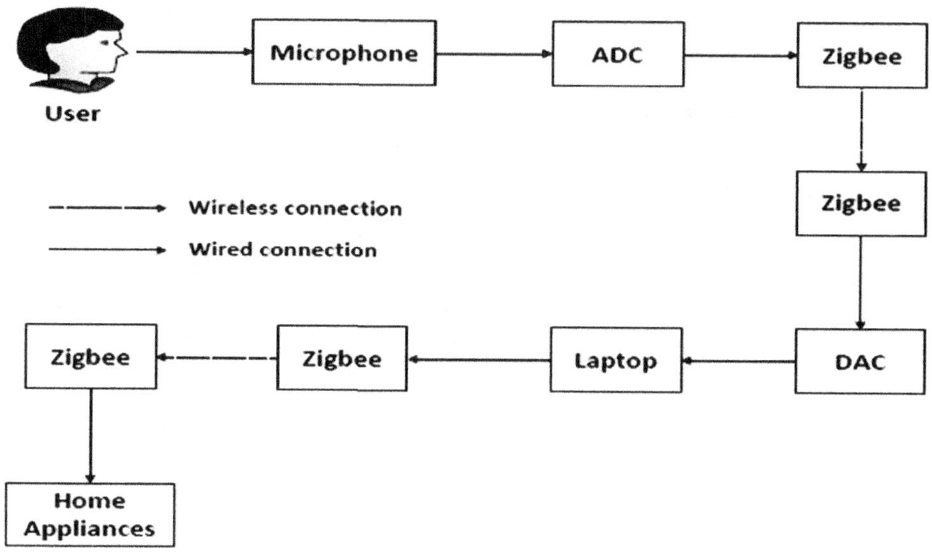

5. Internet of Things (IoT) Based Home Automation System

A home automation system created with the help of Internet of Things (IoT) technology is shown in a research given in (Huang et al., 2010). The adaptable and affordable home control and monitoring system uses an embedded micro-web server with IP connectivity to allow users of android-based smartphones to access and manage home appliances from a distance. The architecture of the system is shown in Figure 6 by the combination of the local environment, the local gateway, and the distant environment.

The function of the remote environment is to allow through a smartphone that supports Wi-Fi, 3G or 4G, and an Android application, authorized users can remotely control and monitor household appliances. Home gateway and hardware interface module combines to form a home environment (Kumar et al., 2014) ;(Teymourzadeh et al., 2013);(Pramanik et al., 2016);(Sen et al., 2015);(Sangeetha et al ., 2015). Arduino Ethernet shield is used to build a key element of home gateway, which is a micro web server. Wired connection is set up to interface hardware modules with actuators and sensors. Devices including power outlets, light switches, door locks, and heating, ventilation, and air conditioning (HVAC) are incorporated to show the control system's viability and efficacy. Home environments support current, human, and temperature sensors for the monitoring system.

Figure 7. Block diagram for IoT based HAS

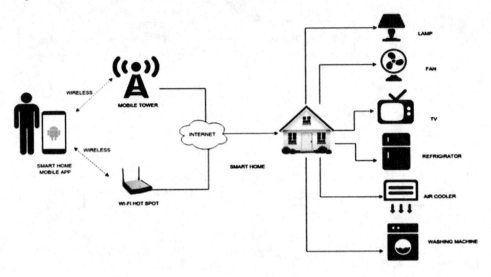

Advantage: In contrast to comparable systems, this system includes a revolutionary communication protocol that allows it to monitor and control the home environment in addition to switching capabilities without the need for a dedicated server PC.

COMPONENTS REQUIRED

1. Arduino UNO

Which can interact with various expansion circuits. This is the ATmega328P microcontroller. It has 14 digital input/output pins, of which 6 pins can be used as PWM (pulse width modulation) outputs, 6 pins of analogue input, a 16 MHz ceramic resonator, a USB connection, a reset button, An ICSP header, and a power jack. It has everything needed to power up the microcontroller and connects it to a PC or laptop with a USB cable or an AC to DC adapter or battery to power it up. Users can experiment with their UNO without worrying too much about doing anything about making a mistake. Its interface between the hardware and the software part, the Bluetooth module transmits the text commands to the Arduino serial port, shown in Figure 8. Similar to various combinations of pre-defined text commands to turn devices on or off. The device name and command for on/off are stored as predefined. For example, if a user wants to turn off a fan, the user must say "fan off" and if he wants to turn on the fan, he must say "fanon". The devices are connected by a relay board to Arduino UNO pins. When the same text command is detected,

Figure 8. Arduino UNO

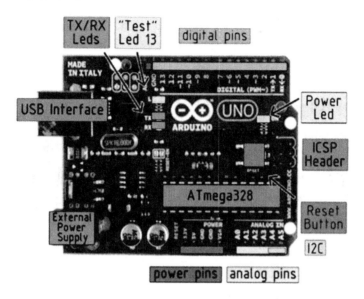

different pin numbers are assigned to the HIGH or LOW output signals to turn the equipment on/off respectively.

2. HC-05 Bluetooth Module

The HC-05 Bluetooth module is becoming popular these days as it is also easily available in electronic shops, shown in Figure 9. It is user friendly and importantly compatible with Arduinos. The Bluetooth module HC-05 is within a range of 10 meters, and its transfer rate of data can vary up to 1Mbps. Furthermore, it provides a switching mode between slave and master mode which means that it is capable of neither transmitting nor receiving data. it has 6 pins here pin 1 is "state", it acts as a status indicator, when the module is not connected or paired with any other device the signal goes low and it is low state That the LED is flashing continuously that the module is not paired with any device. When the module is connected to or paired with another Bluetooth device, the signal goes to HIGH in this high state, the LED blinks continuously with a delay. Which indicates that the module is coupled. Pin 2 and Pin 3 are "Rx and Tx", these two pins act as interface communication. Bluetooth uses different methods for their communication. Rx (receiver) means to receive the signal. Tx (transmitter) means transmit signal then pin 5 is "Vcc" where the module can be operated within 4V to 6V of the power supply. Pin 4 is "Ground" and the last pin 6 is "Key/N" in some modules you will find enable instead of a key,

Figure 9. HC-05 Bluetooth Module

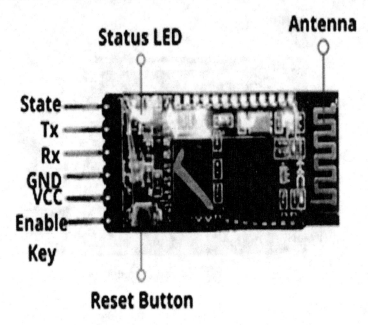

so here they are both the same when the key pin is set to HIGH, so then this module command can work as a mod. Otherwise, it is in data mode by default.

3. 5V Relay Module

The 5v relay module is a 4-channel relay, shown in Figure 10. This is an interface board, and all 4 channels here require a 15-20mA driver current. It can manage various devices with a large amount of current. It is also equipped with a high current relay - also, its work under AC250V 10A or DC30V 10A. It can be controlled directly by the microcontroller.

4. Capacitive Touch Sensor

A capacitive touch sensor is a tactile sensor material that does not require any force to activate the devices, shown in Figure 11. The most surprising feature of the capacitive touch sensor is its ability to feel through its completely covered housing. It works by detecting a change of capacitance due to the impact of an external object, and capacitive sensors may include electronic circuits that measure capacitance across electrodes. Also, our experience is that users do not need to be electrically connected to the circuit for the sensor to operate. The capacitance of a capacitive

Figure 10. 5V Relay Module

touch sensor depends on how close our hand is to the plate of the capacitor, and each touch sensor requires only one wire connected to it. In addition, it is hidden under any non-metallic material and is conveniently used in place of a switch or button. It detects a human hand from a distance of a few inches when needed.

Figure 11. Capacitive touch sensor

5. Android App

Android is open-source software. It can work independently across devices. Its source code is Android Open-Source Article (AOSP). It is based on the Linux operating system (OS) for mobile devices such as tablets and smartphones. Android is developed by Open Handset Alliance and commercially sponsored by Google. The public beta Android version 1.0 launched for developers in "November 2007", with the first commercial Android devices launched in "September 2008". Java has been the default development language for writing Android apps as well. It is the most popular language for Android development. Once Android is developed applications for Android can be easily packaged and also sold through stores like Google Play, Opera Mobile Store, SlideMe, Amazon App Store, etc.

6. Arduino Bluetooth Controller- All in One

Arduino Bluetooth controller all in one application is used in this system. This enables our device to be remotely controlled with the Bluetooth module and Arduino UNO board, shown in Figure 12. It is best to control any electrical appliance in different ways. It can connect and control any Arduinos based device so easily.

METHODOLOGY

The voice-activated and smart touch switch-based home automation system makes our home more comfortable, customized and secure. Voice and button commands are given to the microcontroller via the HC-05 module from the smartphone. The Bluetooth module HC-05 Rx and Tx pins are connected to the microcontroller Tx and Rx pins. The ADC pin of the microcontroller is used as a capacitive touch button to deliver touch commands. The microcontroller processes the commands; then, it sent a command to the relay module as programmed- the relay module acts as a switch for the equipment. The microcontroller is programmed to provide a logic signal output to turn on/off our devices according to the desired inputs, and block diagram of system is shown in Figure 13. This is a block diagram of voice-activated and smart touch switch-based home automation.

IMPLEMENTATION

In this chapter, we can easily implement this system using the above component. First, we need to attach Bluetooth module HC-05, relay module, capacitive touch sensor with Arduino, then turn on supply. After that, start the "Arduino Bluetooth Controller" application on your smartphone. Using the Arduino Bluetooth Controller application, we can easily turn on and off our devices by just giving voice commands. The user gives commands to the microcontroller through the buttons provided in the application. When the microcontroller receives instructions from the Bluetooth module HC-05, it will signal the relay module to turn on/off the devices accordingly. First of all, we need to pair our smartphone with Bluetooth module HC-05 using smartphone Bluetooth setting and search for our Bluetooth module, and we need to enter 4-digit pairing code to pair the device. Once it is paired, you need to start the application. It will automatically select our Bluetooth module and start the application. When we use the voice control option, the application reads our voice, converts our voice into text, and then transmits this text to the microcontroller via Bluetooth module HC-05, and a port in serial mode is used for this purpose. After that, the microcontroller reads the data, decodes the input value, and then sends a signal to the parallel port to turn on the relay module and turn on/off the special equipment. Similarly, when we use the app button option. The application generates a special text that is assigned to each button, then this text is transmitted to the microcontroller via the Bluetooth module HC-05, and the port in serial mode is used for this purpose; After that the microcontroller reads the data and decodes the input value and then sends a signal to the parallel port to turn on the relay module and switch on/off the particular equipment. When we use the capacitive touch button. If the capacitive sensor is not touched, the sensor has no effective capacitance. The voltage dropped across the sensitivity resistor does not affect the other terminal of the sensor. When a finger touches the sensor, the sensor has an effective capacitance, and due to this capacitance, a voltage is induced at the output terminal of the sensor. The voltage is read by the microcontroller pins and converted into digital readings. The Arduino UNO has a 10-bit ADC channel, so the maximum value of a digitized reading is one thousand. The reference value 500 is taken for voltage comparison. The Arduino code detects touch by comparing the voltage reading with a reference value and the position of each device by a set of variables. Upon tapping a sensor, the device concerned is triggered either from the ON position to the OFF position if it is on or from the OFF position to the ON position if it is OFF. Arduino board has logic set to set logic low to the corresponding pin connecting the device through the relay circuit to turn the appliance on and the corresponding pin to turn the appliance off. When a logic high is given to the pin connecting the relay circuit, it switches on the equipment, and when the pin connecting the relay circuit is given a logic low, it turns off the equipment, shown in Figure 14.

Advantages of Arduino Based System for home automation applications(Pramanik et al., 2016); (Sen et al., 2015);(Sangeetha et al ., 2015);(Baviskar et al.,2015);(Piyare et al., 2013).

Figure 12. Android App (Android Bluetooth Controller-All in One)

Figure 13. Block Diagram of home automation system

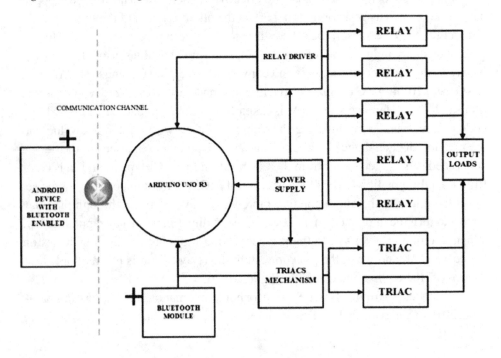

Figure 14. Prototype of Home Automation System Controlled By Using Mobile And Capacitive Touch Through Arduino Based System.

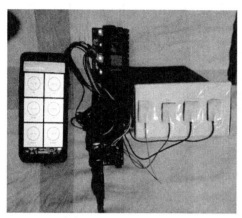 

1. Everything is automated so it is easy to use.
2. It is controlled by a mobile application so no extra training is required.
3. We can change the controlling system as our requirement.
4. It works on Arduino based system so we can easily understand how it works.
5. It saves our time.
6. Every home appliance can be controlled by one android application.
7. Easy installation and user friendly
8. The smart touch switch is used in the system which makes the system shockproof.
9. It replaces the normal switchboard.

CONCLUSION

Home automation can be concluded from the above discussion that the voice-activated and smart touch switch-based home automation system is an advantageous and high-tech based device. This device can operate home appliances and provide multiple ways to control them as the system operates in three ways which are very useful and give users an efficient way to control appliances by their voice and button convenience provide. The voice-activated and smart touch switch-based home automation device is smaller in size, requires less space, and is easy to set up. It is a low-cost system due to the material used in it. Using this system allows users to control the number of devices, and it also minimizes electrical hazards and system flow chart.

REFERENCES

Amoran, A. E., Oluwole, A. S., Fagorola, E. O., & Diarah, R. S. (2021). Home automated system using Bluetooth and an android application. *Scientific American*, *11*, e00711.

Baviskar, J., Mulla, A., Upadhye, M., Desai, J., & Bhovad, A. (2015, January). Performance analysis of ZigBee based real time Home Automation system. In *2015 International Conference on Communication, Information & Computing Technology (ICCICT)* (pp. 1-6). IEEE.

Budiharto, W., Gunawan, A. A. S., Irwansyah, E., & Suroso, J. S. (2019, December). Android-based wireless controller for military robot using bluetooth technology. In *2019 2nd World Symposium on Communication Engineering (WSCE)* (pp. 215-219). IEEE.

Chinchansure, P. S., & Kulkarni, C. V. (2014, January). Home automation system based on FPGA and GSM. In *2014 International Conference on Computer Communication and Informatics* (pp. 1-5). IEEE.

Felix, C., & Raglend, I. J. (2011, July). Home automation using GSM. In *2011 International Conference on Signal Processing, Communication, Computing and Networking Technologies* (pp. 15-19). IEEE.

Huang, H., Xiao, S., Meng, X., & Xiong, Y. (2010, April). A remote home security system based on wireless sensor network and GSM technology. In *2010 Second International Conference on Networks Security, Wireless Communications and Trusted Computing* (Vol. 1, pp. 535-538). IEEE.

Ishak, M. H. I. (2014). Bluetooth-based home automation system using an android phone. *Jurnal Teknologi, 70*(3).

Kannapiran, S., & Chakrapani, A. (2017). A novel home automation system using Bluetooth and Arduino. *International Journal of Advances in Computer and Electronics Engineering, 2*(2), 41-44.

Kumar, S., & Lee, S. R. (2014, June). Android based smart home system with control via Bluetooth and internet connectivity. In *The 18th IEEE International Symposium on Consumer Electronics (ISCE 2014)* (pp. 1-2). IEEE.

Pandya, B., Mehta, M., Jain, N., & Kadam, S. (2016). Android Based Home Automation System Using Bluetooth & Voice Command–Implementation. *International Research Journal of Engineering and Technology*.

Patel, H. K., Gaur, V., Kumar, S., Singh, A. K., & Mittal, S. (2021). Review on Next Step Home Automation Using Wi-Fi Module. In *Advances in Smart Communication and Imaging Systems* (pp. 301–307). Springer.

Piyare, R. (2013). Internet of things: ubiquitous home control and monitoring system using android based smart phone. *International Journal of Internet of Things, 2*(1), 5-11.

Piyare, R., & Tazil, M. (2011, June). Bluetooth based home automation system using cell phone. In *2011 IEEE 15th International Symposium on Consumer Electronics (ISCE)* (pp. 192-195). IEEE.

Pramanik, A., Nagar, V., Dwivedi, S., & Choudhury, B. (2016, March). GSM based Smart home and digital notice board. In *2016 International Conference on Computational Techniques in Information and Communication Technologies (ICCTICT)* (pp. 41-46). IEEE.

Raheem, A. K. K. A. (2017). Bluetooth based smart home automation system using arduino UNO microcontroller. *Al-Mansour Journal, 27*, 139–156.

Ramlee, R. A., Othman, M. A., Leong, M. H., Ismail, M. M., & Ranjit, S. S. S. (2013, March). Smart home system using android application. In *2013 International Conference of Information and Communication Technology (ICoICT)* (pp. 277-280). IEEE.

Ransing, R. S., & Rajput, M. (2015, January). Smart home for elderly care, based on Wireless Sensor Network. In *2015 International Conference on Nascent Technologies in the Engineering Field (ICNTE)* (pp. 1-5). IEEE.

Sangeetha, S. B. (2015, March). Intelligent interface based speech recognition for home automation using android application. In *2015 International Conference on Innovations in Information, Embedded and Communication Systems (ICIIECS)* (pp. 1-11). IEEE.

Sen, S., Chakrabarty, S., Toshniwal, R., & Bhaumik, A. (2015). Design of an intelligent voice controlled home automation system. *International Journal of Computers and Applications, 121*(15).

Teymourzadeh, R., Ahmed, S. A., Chan, K. W., & Hoong, M. V. (2013, December). Smart GSM based home automation system. In *2013 IEEE Conference on Systems, Process & Control (ICSPC)* (pp. 306-309). IEEE.

Ullah, M. A., & Celik, A. R. (2016). An effective approach to build smart building based on Internet of Things (IoT). *Journal of Basic and Applied Scientific Research, 6*, 56–62.

Compilation of References

Thakker, S., & Narayanamoorthi, R. (2015). Smart and wireless waste management. *2015 International Conference on Innovations in Information, Embedded and Communication Systems (ICIIECS).* 10.1109/ICIIECS.2015.7193141

Kumari, P. (2018, January 23). *Iot based smart waste bin model to optimize the waste management process.* http://dl.lib.uom.lk/handle/123/13033

Ramson, S. J., & Moni, D. J. (2017). Wireless sensor networks based Smart-Bin. *Computers &amp. Electrical Engineering, 64,* 337–353. doi:10.1016/j.compeleceng.2016.11.030

Ravi, V. R., Hema, M., SreePrashanthini, S., & Sruthi, V. (2021). Smart-Bins for trash monitoring in smart cities using IoT system. *IOP Conference Series. Materials Science and Engineering, 1055*(1), 012078. doi:10.1088/1757-899X/1055/1/012078

A Smart Information System for Public Transportation Using IoT. (2017). *International Journal of Recent Trends in Engineering and Research, 3*(4), 222–230. doi:10.23883/IJRTER.2017.3138. YCHJE

A, J., Nagarajan, R., Satheeshkumar, K., Ajithkumar, N., Gopinath, P., & Ranjithkumar, S. (2017). Intelligent Smart Home Automation and Security System Using Arduino and Wi-fi. *International Journal of Engineering And Computer Science.* doi:10.18535/ijecs/v6i3.53

Aadil, F., Ahsan, W., Rehman, Z. U., Shah, P. A., Rho, S., & Mehmood, I. (2018). Clustering algorithm for internet of vehicles (IoV) based on dragonfly optimizer (CAVDO). *The Journal of Supercomputing, 74*(9), 4542–4567. doi:10.100711227-018-2305-x

Abbas, M. T., Muhammad, A., & Song, W. C. (2019). SD-IoV: SDN enabled routing for internet of vehicles in road-aware approach. *Journal of Ambient Intelligence and Humanized Computing, 11*(3), 1265–1280. doi:10.100712652-019-01319-w

Abduljabbar, R., Dia, H., Liyanage, S., & Bagloee, S. A. (2019). Applications of Artificial Intelligence in Transport: An Overview. *Sustainability, 11*(1), 189. doi:10.3390u11010189

Abidin, A. F., Kolberg, M., & Hussain, A. (2015). Integrating Twitter Traffic Information with Kalman Filter Models for Public Transportation Vehicle Arrival Time Prediction. *Big-Data Analytics and Cloud Computing,* 67–82. doi:10.1007/978-3-319-25313-8_5

Abubakar, A., Ajuji, M., & Usman Yahya, I. (2020). Comparison of deep transfer learning techniques in human skin burns discrimination. *Applied System Innovation*, *3*(2).

Ahmed, M. S. (2019). Technical Skill Assessment using Machine Learning and Artificial Intelligence Algorithm. *International Journal of Engine Research*, *8*(12). Advance online publication. doi:10.17577/IJERTV8IS120109

Ahmood, Z. (2020). *Connected Vehicles in the Internet of Things*. Springer Nature Switzerland AG. doi:10.1007/978-3-030-36167-9

Alaa, M., Zaidan, A., Zaidan, B., Talal, M., & Kiah, M. (2017). A review of smart home applications based on Internet of Things. *Journal of Network and Computer Applications*, *97*, 48–65. doi:10.1016/j.jnca.2017.08.017

Al-Ali, A., Zualkernan, I. A., Rashid, M., Gupta, R., & Alikarar, M. (2017). A smart home energy management system using IoT and big data analytics approach. *IEEE Transactions on Consumer Electronics*, *63*(4), 426–434. doi:10.1109/TCE.2017.015014

Al-Ghaili, A. M., Kasim, H., Othman, M., & Hashim, W. (2020). QR code based authentication method for IoT applications using three security layers. *TELKOMNIKA*, *18*(4), 2004. doi:10.12928/telkomnika.v18i4.14748

Al-Hassan, E., Shareef, H., Islam, M. M., Wahyudie, A., & Abdrabou, A. A. (2018). Improved Smart Power Socket for Monitoring and Controlling Electrical Home Appliances. *IEEE Access: Practical Innovations, Open Solutions*, *6*, 49292–49305. doi:10.1109/ACCESS.2018.2868788

Ali Hasnain, H. S. A. (2021). Agriculture Monitoring System Using IoT: A Review Paper. *International Journal on Recent and Innovation Trends in Computing and Communication, 9*(1), 1–6. doi:10.17762/ijritcc.v9i1.5452

Ali, M. L., Alam, M., & Rahaman, M. A. N. R. (2012). RFID-based e-monitoring system for municipal solid waste management. *2012 7th International Conference on Electrical and Computer Engineering*. 10.1109/ICECE.2012.6471590

Ali, A. H., Chisab, R. F., & Mnati, M. J. (2019). A smart monitoring and controlling for agricultural pumps using LoRa IOT technology. *Indonesian Journal of Electrical Engineering and Computer Science*, *13*(1), 286. doi:10.11591/ijeecs.v13.i1.pp286-292

Almusaylim, A. Z., & Jhanjhi, N. (2020). Comprehensive Review: Privacy Protection of User in Location-Aware Services of Mobile Cloud Computing. *Wireless Personal Communications*, *111*, 541–564. doi:10.100711277-019-06872-3

Amoran, A. E., Oluwole, A. S., Fagorola, E. O., & Diarah, R. S. (2021). Home automated system using Bluetooth and an android application. *Scientific American*, *11*, e00711.

Amruta, B. P., Madhuri, S., Pallavi, P., & Satish, S. (2018). City Garbage Collection Indicator Using Wireless Communication. *International Journal of Scientific Research in Science, Engineering and Technology*, *4*, 378–380.

Anjali, Khangar, & Bhakre. (2018). A review paper on effective agriculture monitoring system using IoT. *International Journal of Modern Trends in Engineering & Research, 5*(3), 15–17. doi:10.21884/IJMTER.2018.5058.YLOGA

Arefnezhad, S., Hamet, J., Eichberger, A., Frühwirth, M., Ischebeck, A., Koglbauer, I. V., Moser, M., & Yousefi, A. (2022). Driver drowsiness estimation using EEG signals with a dynamical encoder–decoder modeling framework. *Scientific Reports, 12*(1), 1–18. doi:10.103841598-022-05810-x PMID:35173189

Asnafi, M., & Dastgheibifard, S. (2018, July 9). A Review on Potential Applications of Unmanned Aerial Vehicle for Construction Industry. *Sustainable Structures and Materials. International Journal (Toronto, Ont.), 1*(2), 44–53.

Ata, A., Khan, M. A., Abbas, S., Ahmad, G., & Fatima, A. (2019). Modelling smart road traffic congestion control system using machine learning techniques. *Neural Network World, 29*(2), 99–110. doi:10.14311/NNW.2019.29.008

Automazione, D. (2016). *GI dataset.* https://www.deltamaxautomazione.it/risolvi/

Avasthi, S., Sanwal, T., Sareen, P., & Tripathi, S. L. (2022). Augmenting Mental Healthcare with Artificial Intelligence, Machine Learning, and Challenges in Telemedicine. In Handbook of Research on Lifestyle Sustainability and Management Solutions Using AI, Big Data Analytics, and Visualization (pp. 75-90). IGI Global.

Avasthi, S., Chauhan, R., & Acharjya, D. P. (2022). Topic Modeling Techniques for Text Mining Over a Large-Scale Scientific and Biomedical Text Corpus. *International Journal of Ambient Computing and Intelligence, 13*(1), 1–18. doi:10.4018/IJACI.293137

Bah, M. D., Hafiane, A., & Canals, R. (2017). Weeds detection in UAV imagery using SLIC and the hough transform. *2017 Seventh International Conference on Image Processing Theory, Tools and Applications (IPTA).* 10.1109/IPTA.2017.8310102

Balakrishnan, V., Varadharajan, V., Tupakula, U., & Lues, P. (2007). Team: Trust enhanced security architecture for mobile ad-hoc networks. *15th IEEE International Conference on Networks ICON 2007,* 182–187. 10.1109/ICON.2007.4444083

Baldessari, R., Bödekker, B., Deegener, M., Festag, A., Franz, W., Kellum, C. C., ... Zhang, W. (2007). *Car-2-car communication consortium-manifesto.* Academic Press.

Baldo, D., Mecocci, A., Parrino, S., Peruzzi, G., & Pozzebon, A. (2021). A Multi-Layer LoRaWAN Infrastructure for Smart Waste Management. *Sensors (Basel), 21*(8), 2600. doi:10.339021082600 PMID:33917255

Bano, A. (2020, December 29). *AIoT-Based Smart Bin for Real-Time Monitoring and Management of Solid Waste.* Https://Www.Hindawi.Com/Journals/Sp/2020/6613263/. https://www.hindawi.com/journals/sp/2020/6613263/

Bansod, B., Singh, R., Thakur, R., & Singhal, G. (2017). A comparision between satellite based and drone based remote sensing technology to achieve sustainable development: A review. *Journal of Agriculture and Environment for International Development, 111*(2), 383–407. doi:10.12895/jaeid.20172.690

Bauza, R., Gozalvez, J., & Sanchez-Soriano, J. (2010). Road traffic congestion detection through cooperative Vehicle-to-Vehicle communications. *Proceedings - Conference on Local Computer Networks*, 606–612. 10.1109/LCN.2010.5735780

Baviskar, J., Mulla, A., Upadhye, M., Desai, J., & Bhovad, A. (2015, January). Performance analysis of ZigBee based real time Home Automation system. In *2015 International Conference on Communication, Information & Computing Technology (ICCICT)* (pp. 1-6). IEEE.

Belhadi, A., Mani, V., Kamble, S. S., Kah, S. A. R., & Verma, S. (n.d.). *Artificial intelligence-driven innovation for enhancing supply chain resilience and performance under the effect of supply chain dynamism: an empirical investigation.* Academic Press.

Beretas, C. (2019). Internet of Things (IOT) is Smart Homes and the Risks. *Journal of Current Engineering and Technology, 1*(3), 1–2. doi:10.36266/JCET/115

Boukerche, A., Tao, Y., & Sun, P. (2020). Artificial intelligence-based vehicular traffic flow prediction methods for supporting intelligent transportation systems. *Computer Networks, 182*, 107484. doi:10.1016/j.comnet.2020.107484

Budiharto, W., Gunawan, A. A. S., Irwansyah, E., & Suroso, J. S. (2019, December). Android-based wireless controller for military robot using bluetooth technology. In *2019 2nd World Symposium on Communication Engineering (WSCE)* (pp. 215-219). IEEE.

Bui, K. H. N., Jung, J. E., & Camacho, D. (2017). Game theoretic approach on Real-time decision making for IoT-based traffic light control. *Concurrency and Computation, 29*(11), e4077. doi:10.1002/cpe.4077

Camp, T., Boleng, J., & Davies, V. (n.d.). A survey of mobility models for ad hoc network research. *Wirel. Commun. Mob. Comput., 2*, 483–502. https://online library.wiley.com/doi/10.1002/wcm.72

Canny, J. (1986). A Computational Approach to Edge Detection. *IEEE Transactions on Pattern Analysis and Machine Intelligence, PAMI-8*(6), 679–698. doi:10.1109/TPAMI.1986.4767851 PMID:21869365

CGS. (n.d.). *Top COVID Challenges for Contact Centers in 2021.* Retrieved March 12, 2022, from https://www.cgsinc.com/blog/top-covid-challenges-contact-centers-2021

Chauhan, R., Avasthi, S., Alankar, B., & Kaur, H. (2021). Smart IoT Systems: Data Analytics, Secure Smart Home, and Challenges. In Transforming the Internet of Things for Next-Generation Smart Systems (pp. 100-119). IGI Global.

Cha, Y.-J., Choi, W., Suh, G., Mahmoudkhani, S., & Buyukozturk, O. (2017). Autonomous structural visual inspection using region-based deep learning for detecting multiple damage types. *Computer-Aided Civil and Infrastructure Engineering, 33*(9), 731–747. doi:10.1111/mice.12334

Chen, L. B., Su, K. Y., Mo, Y. C., Chang, W. J., Hu, W. W., Tang, J. J., & Yu, C. T. (2018, September). An Implementation of Deep Learning based IoV System for Traffic Accident Collisions Detection with an Emergency Alert Mechanism. *2018 IEEE 8th International Conference on Consumer Electronics - Berlin (ICCE-Berlin).* 10.1109/ICCE-Berlin.2018.8576197

Chen, C. J., Huang, Y. Y., Li, Y. S., Chang, C. Y., & Huang, Y. M. (2020). An AIoT Based Smart Agricultural System for Pests Detection. *IEEE Access: Practical Innovations, Open Solutions, 8*, 180750–180761. doi:10.1109/ACCESS.2020.3024891

Chen, S. H., & Perng, D. B. (2011). Directional textures auto-inspection using principal component analysis. *International Journal of Advanced Manufacturing Technology, 55*(9-12), 1099–1110. doi:10.100700170-010-3141-1

Chen, W. (2015). *A book on Vehicular Communications and Networks Architectures, Protocols, Operations and Deployment.* Elsevier.

Chen, W.-C., & Hsu, S.-W. (2007). A neural-network approach for an automatic LED inspection system. *Expert Systems with Applications, 33*(2), 531–537. doi:10.1016/j.eswa.2006.06.011

Chen, W., Gao, Y., Gao, L., & Li, X. (2018). A new ensemble approach based on deep convolutional neural networks for steel surface defect classification. *Procedia CIRP, 72*, 1069–1072. doi:10.1016/j.procir.2018.03.264

Chen, W., Guha, R. K., Kwon, T. J., Lee, J., & Hsu, Y. Y. (2011). A survey and challenges in routing and data dissemination in vehicular ad hoc networks. *Wireless Communications and Mobile Computing, 11*(7), 787–795. doi:10.1002/wcm.862

Chettri, L., & Bera, R. (2020). A Comprehensive Survey on Internet of Things (IoT) Toward 5G Wireless Systems. *IEEE Internet of Things Journal, 7*(1), 16–32. doi:10.1109/JIOT.2019.2948888

Chinchansure, P. S., & Kulkarni, C. V. (2014, January). Home automation system based on FPGA and GSM. In *2014 International Conference on Computer Communication and Informatics* (pp. 1-5). IEEE.

Chirra, V. R. R., ReddyUyyala, S., & Kolli, V. K. K. (2019). Deep CNN: A Machine Learning Approach for Driver Drowsiness Detection Based on Eye State. *Rev. d'Intelligence Artif., 33*(6), 461-466.

Chollet, F. (2021). *Image classification from scratch.* https://keras.io/examples/vision/image_classification_from_scratch/

Chollet, F. (2017). *Deep Learning with Python* (1st ed.). Manning Publications Co.

Chowdhury, M. M., Hasan, M., Safait, S., Chaki, D., & Uddin, J. (2018, June). A Traffic Congestion Forecasting Model using CMTF and Machine Learning. *2018 Joint 7th International Conference on Informatics, Electronics & Vision (ICIEV) and 2018 2nd International Conference on Imaging, Vision & Pattern Recognition (icIVPR).* 10.1109/ICIEV.2018.8640985

Christina. (2020). *An Introduction to Logistic Regression for Categorical Data Analysis.* Retrieved March 12, 2022, from https://towardsdatascience.com/an-introduction-to-logistic-regression-for-categorical-data-analysis-7cabc551546c

Cognitive Automation Community. (2020). *5 Ways Cognitive Automation is Transforming Supply Chain Processes.* Retrieved March 5, 2022, from https://www.cognitiveautomation.com/resources/5-ways-cognitive-automation-is-transforming-supply-chain-processes

Contractor, S. (2020). *Coronavirus pandemic: Nissan temporarily halts operations in Asia, Africa, Middle East.* Carandbike. Available: https://auto.ndtv.com/news/coronavirus-pandemic-nissan-temporarily-haltsoperations-in-asia-africa-middle-east-2200191

Da Cunha, F. D., Boukerche, A., Villas, L., Viana, A. C., & Loureiro, A. A. (2014). *Data communication in VANETs: a Survey, Challenges and Applications* (Research Report RR-8498). INRIA. https://hal.inria.fr/hal-00981126/ document

Dadshani, S., Kurakin, A., Amanov, S., Hein, B., Rongen, H., Cranstone, S., Blievernicht, U., Menzel, E., Léon, J., Klein, N., & Ballvora, A. (2015). Non-invasive assessment of leaf water status using a dual-mode microwave resonator. *Plant Methods, 11*(1), 8. doi:10.118613007-015-0054-x PMID:25918549

Dandala, T. T., Krishnamurthy, V., & Alwan, R. (2017, January). Internet of Vehicles (IoV) for traffic management. *2017 International Conference on Computer, Communication and Signal Processing (ICCCSP).* 10.1109/ICCCSP.2017.7944096

Daraghmi, Y. A., Yi, C. W., & Stojmenovic, I. (2013). Forwarding methods in data dissemination and routing protocols for vehicular ad hoc networks. *IEEE Network, 27*(6), 74–79. doi:10.1109/MNET.2013.6678930

Davis, S. C., Diegel, S. W., & Boundy, R. G. (2018). *Transportation Energy Data Book* (36th ed.). Oak Ridge National Laboratory.

Dayalani, V. (2021). *12 Charts That Show The Rise Of Indian Tech During Covid.* Retrieved October 13, 2001, from https://inc42.com/datalab/12-charts-that-show-the-rise-of-indian-tech-after lockdown/

De, S., Mukherjee, A., & Ullah, E. (2018). *Convergence guarantees for RMSProp and ADAM in non-convex optimization and an empirical comparison to Nesterov acceleration.* https://arxiv.org/abs/1807.06766

Deloitte. (2022). *Understanding COVID-19's impact on the automotive sector.* Retrieved April 25, 2022, from https://www2.deloitte.com/us/en/pages/about-deloitte/articles/covid-19/covid-19-impact-on-automotive-sector.html

Deng, J., Dong, W., Socher, R., Li, L., Li, K., & Li, F.-F. (2009). ImageNet: A large-scale hierarchical image database. *2009 IEEE Conference on Computer Vision and Pattern Recognition,* 248–255. 10.1109/CVPR.2009.5206848

Dinh, T., Kim, Y., & Lee, H. (2017). A Location-Based Interactive Model of Internet of Things and Cloud (IoT-Cloud) for Mobile Cloud Computing Applications. *Sensors (Basel)*, *17*(3), 489. doi:10.339017030489 PMID:28257067

Dressler, F., Kargl, F., Ott, J., Tonguz, O. K., & Wischhof, L. (2011). Research challenges in intervehicular communication: Lessons of the 2010 Dagstuhl Seminar. *IEEE Communications Magazine*, *49*(5), 158–164. doi:10.1109/MCOM.2011.5762813

Dua, M., Singla, R., Raj, S., & Jangra, A. (2021). Deep CNN models-based ensemble approach to driver drowsiness detection. *Neural Computing & Applications*, *33*(8), 3155–3168. doi:10.100700521-020-05209-7

Dubey, S. R., Singh, S. K., & Chaudhuri, B. B. (2021). *A Comprehensive Survey and Performance Analysis of Activation Functions in Deep Learning.* arXiv preprint arXiv:2109.14545.

Dubey, R., Gunasekaran, A., Childe, S. J., Bryde, D. J., Giannakis, M., Foropon, C., Roubaud, D., & Hazen, B. T. (2020). Big data analytics and artificial intelligence pathway to operational performance under entrepreneurial orientation and environmental dynamism: A study of manufacturing organizations. *International Journal of Production Economics*, *226*, 107599. doi:10.1016/j.ijpe.2019.107599

Ertuğrul, Ö. F. (2018). A novel type of activation function in artificial neural networks: Trained activation function. *Neural Networks*, *99*, 148–157. doi:10.1016/j.neunet.2018.01.007 PMID:29427841

Essien, A., Petrounias, I., Sampaio, P., & Sampaio, S. (2020). A deep-learning model for urban traffic flow prediction with traffic events mined from twitter. *World Wide Web (Bussum)*. Advance online publication. doi:10.100711280-020-00800-3

Eze, Zhang, & Eze. (2016). *Advances in Vehicular Ad-Hoc Networks (VANETs): Challenges and Road-map for Future Development.* Academic Press.

Faezipour, M., Nourani, M., Saeed, A., & Addepalli, S. (2012). Progress and challenges in intelligent vehicle area networks. *Communications of the ACM*, *55*(2), 90–100. doi:10.1145/2076450.2076470

Fahim, A., Hasan, M., & Chowdhury, M. A. (2021). Smart parking systems: Comprehensive review based on various aspects. *Heliyon*, *7*(5), e07050. doi:10.1016/j.heliyon.2021.e07050 PMID:34041396

Felix, C., & Raglend, I. J. (2011, July). Home automation using GSM. In *2011 International Conference on Signal Processing, Communication, Computing and Networking Technologies* (pp. 15-19). IEEE.

Fontes, F. (2022). *Cognitive Automation Is Building Self-Healing Supply Chains.* Retrieved April 26, 2022, from https://www.supplychainbrain.com/blogs/1-think-tank/post/34424-cognitive-automation-is-building-self-healing-supply-chains

Gajja, M. (2020). Brain Tumor Detection Using Mask R-CNN. *Journal of Advanced Research in Dynamical and Control Systems*, *12*(SP8), 101–108. doi:10.5373/JARDCS/V12SP8/20202506

Gao, D., Sun, Q., Hu, B., & Zhang, S. (2020). A Framework for Agricultural Pest and Disease Monitoring Based on Internet-of-Things and Unmanned Aerial Vehicles. *Sensors (Basel)*, *20*(5), 1487. doi:10.339020051487 PMID:32182732

Gao, D., Zhang, S., Zhang, F., He, T., & Zhang, J. (2019). RowBee: A Routing Protocol Based on Cross-Technology Communication for Energy-Harvesting Wireless Sensor Networks. *IEEE Access: Practical Innovations, Open Solutions*, *7*, 40663–40673. doi:10.1109/ACCESS.2019.2902902

García, L., Parra, L., Jimenez, J. M., Lloret, J., & Lorenz, P. (2020). IoT-Based Smart Irrigation Systems: An Overview on the Recent Trends on Sensors and IoT Systems for Irrigation in Precision Agriculture. *Sensors (Basel)*, *20*(4), 1042. doi:10.339020041042 PMID:32075172

Ghorpade, D. D., & Patki, A. M. (2016). A Review on IOT Based Smart Home Automation Using Renewable Energy Sources. *International Journal of Scientific Research*, *5*(5), 2000–2001. doi:10.21275/v5i5.NOV163878

Gladence, L. M., Anu, V. M., Rathna, R., & Brumancia, E. (2020). Recommender system for home automation using IoT and artificial intelligence. *Journal of Ambient Intelligence and Humanized Computing*. Advance online publication. doi:10.100712652-020-01968-2

Goudarzi, S., Kama, M., Anisi, M., Soleymani, S., & Doctor, F. (2018). Self-Organizing Traffic Flow Prediction with an Optimized Deep Belief Network for Internet of Vehicles. *Sensors (Basel)*, *18*(10), 3459. doi:10.339018103459 PMID:30326567

Grover, J., Gaur, M. S., & Laxmi, V. (2013). Trust establishment techniques in VANET. In *Wireless Networks and Security* (pp. 273–301). Springer. doi:10.1007/978-3-642-36169-2_8

Grover, P., Kar, A. K., & Dwivedi, Y. K. (2020). Understanding artificial intelligence adoption in operations management: Insights from the review of academic literature and social media discussions. *Annals of Operations Research*. Advance online publication. doi:10.100710479-020-03683-9

Gunawan, T. S., Yaldi, I. R. H., Kartiwi, M., Ismail, N., Za'bah, N. F., Mansor, H., & Nordin, A. N. (2017). Prototype Design of Smart Home System using Internet of Things. *Indonesian Journal of Electrical Engineering and Computer Science*, *7*(1), 107. doi:10.11591/ijeecs.v7.i1.pp107-115

Gupta, M., Abdelsalam, M., Khorsandroo, S., & Mittal, S. (2020). Security and Privacy in Smart Farming: Challenges and Opportunities. *IEEE Access: Practical Innovations, Open Solutions*, *8*, 34564–34584. doi:10.1109/ACCESS.2020.2975142

Hafemann, L. G., Oliveira, L. S., Cavalin, P. R., & Sabourin, R. (2015). Transfer learning between texture classification tasks using convolutional neural networks. *2015 International Joint Conference on Neural Networks (IJCNN)*, 1–7. 10.1109/IJCNN.2015.7280558

Haque, K. F. (2020, June 1). *An IoT Based Efficient Waste Collection System with Smart Bins*. IEEE Conference Publication. https://ieeexplore.ieee.org/document/9221251

Haque, K. F., Zabin, R., Yelamarthi, K., Yanambaka, P., & Abdelgawad, A. (2020). An IoT-Based Efficient Waste Collection System with Smart-Bins. *2020 IEEE 6th World Forum on Internet of Things (WF-IoT)*. 10.1109/WF-IoT48130.2020.9221251

Harendra Negi, S. C. (2020). Smart Farming using IoT. *International Journal of Engineering and Advanced Technology*, *8*(4S), 45–51. doi:10.35940/ijeat.D1015.0484S19

Hartenstein, H., & Laberteaux, K. P. (2010). *VANET: Vehicular Applications and Inter-Networking Technologies*. Wiley Online Library. doi:10.1002/9780470740637

Hasan, M. K., Ahsan, M., Newaz, S. H., & Lee, G. M. (2021). Human face detection techniques: A comprehensive review and future research directions. *Electronics (Basel)*, *10*(19), 2354. doi:10.3390/electronics10192354

Hatcher, W. G., & Yu, W. (2018). A Survey of Deep Learning: Platforms, Applications and Emerging Research Trends. *IEEE Access: Practical Innovations, Open Solutions*, *6*, 24411–24432. doi:10.1109/ACCESS.2018.2830661

He, K., Zhang, X., Ren, S., & Sun, J. (2016). Deep residual learning for image recognition. *2016 IEEE Conference on Computer Vision and Pattern Recognition (CVPR)*, 770–778. 10.1109/CVPR.2016.90

Helwan, A., Ma'aitah, M., Abiyev, R., Uzelaltınbulat, S., & Sonyel, B. (2021). Deep learning based on residual networks for automatic sorting of bananas. *Journal of Food Quality*, *2021*, 1–11. doi:10.1155/2021/5516368

Hershey, S., Chaudhuri, S., Ellis, D. P. W., Gemmeke, J. F., Jansen, A., Moore, R. C., Plakal, M., Platt, D., Saurous, R. A., Seybold, B., Slaney, M., Weiss, R. J., & Wilson, K. (2017). CNN architectures for large-scale audio classification. *2017 IEEE International Conference on Acoustics, Speech and Signal Processing (ICASSP)*. 10.1109/ICASSP.2017.7952132

He, W., Yan, G., & Xu, L. D. (2014, May). Developing Vehicular Data Cloud Services in the IoT Environment. *IEEE Transactions on Industrial Informatics*, *10*(2).

Hongyan, K. (2011). Design and Realization of Internet of Things Based on Embedded System Used in Intelligent Campus. *International Journal of Advancements in Computing Technology*, *3*(11), 291–298. doi:10.4156/ijact.vol3.issue11.37

Hossain, E., Chow, G., Leung, V. C. M., McLeod, R. D., Mišić, J., Wong, V. W. S., & Yang, O. (2010). Vehicular telematics over heterogeneous wireless networks: A survey. *Computer Communications*, *33*(7), 775–793. doi:10.1016/j.comcom.2009.12.010

Huang, H., Xiao, S., Meng, X., & Xiong, Y. (2010, April). A remote home security system based on wireless sensor network and GSM technology. In *2010 Second International Conference on Networks Security, Wireless Communications and Trusted Computing* (Vol. 1, pp. 535-538). IEEE.

Hussain, R., Rezaeifar, Z., Lee, Y. H., & Oh, H. (2015). Secure and privacy-aware traffic information as a service in VANET- based clouds. *Pervasive and Mobile Computing*, *24*, 194–209. doi:10.1016/j.pmcj.2015.07.007

Intelligent Transport Systems (ITS). (2012). *Framework for public mobile networks in cooperative its (c-its)s* (Tech. Rep.). European Telecommunications Standards Institute (ETSI).

Ishak, M. H. I. (2014). Bluetooth-based home automation system using an android phone. *Jurnal Teknologi, 70*(3).

Islam, N., Rashid, M. M., Pasandideh, F., Ray, B., Moore, S., & Kadel, R. (2021). A Review of Applications and Communication Technologies for Internet of Things (IoT) and Unmanned Aerial Vehicle (UAV) Based Sustainable Smart Farming. *Sustainability, 13*(4), 1821. doi:10.3390u13041821

Islam, S., Sithamparanathan, K., Chavez, K. G., Scott, J., & Eltom, H. (2019). Energy efficient and delay aware ternary-state transceivers for aerial base stations. *Digital Communications and Networks, 5*(1), 40–50. doi:10.1016/j.dcan.2018.10.007

Jadon, S., Choudhary, A., Saini, H., Dua, U., Sharma, N., & Kaushik, I. (2020). Comfy Smart Home using IoT. SSRN *Electronic Journal.* doi:10.2139/ssrn.3565908

Jayaraman, P., Yavari, A., Georgakopoulos, D., Morshed, A., & Zaslavsky, A. (2016). Internet of Things Platform for Smart Farming: Experiences and Lessons Learnt. *Sensors (Basel), 16*(11), 1884. doi:10.339016111884 PMID:27834862

Jha, K., Doshi, A., Patel, P., & Shah, M. (2019). A comprehensive review on automation in agriculture using artificial intelligence. *Artificial Intelligence in Agriculture, 2*, 1–12. doi:10.1016/j.aiia.2019.05.004

Jog, Y., Sajeev, A., Vidwans, S., & Mallick, C. (2015). Understanding Smart and Automated Parking Technology. *International Journal of U- and e-Service Science and Technology, 8*(2), 251–262. doi:10.14257/ijunesst.2015.8.2.25

John, J., Varkey, M. S., Podder, R. S., Sensarma, N., Selvi, M., Santhosh Kumar, S. V. N., & Kannan, A. (2021). Smart Prediction and Monitoring of Waste Disposal System Using IoT and Cloud for IoT Based Smart Cities. *Wireless Personal Communications, 122*(1), 243–275. doi:10.100711277-021-08897-z

Jusat, N., Zainuddin, A. A., Sahak, R., Andrew, A. B., Subramaniam, K., & Rahman, N. A. (2021, August). Critical Review In Smart Car Parking Management Systems. *2021 IEEE 7th International Conference on Smart Instrumentation, Measurement and Applications (ICSIMA).* 10.1109/ICSIMA50015.2021.9526322

K, A., & S, N. S. R. (2019). Analysis of Machine Learning Algorithm in IOT Security Issues and Challenges. *Journal of Advanced Research in Dynamical and Control Systems, 11*(9), 1030–1034. doi:10.5373/JARDCS/V11/20192668

K., D. S. (2020). IoT based Smart Energy Theft Detection System in Smart Home. *Journal of Advanced Research in Dynamical and Control Systems, 12*(SP8), 605–613. doi:10.5373/JARDCS/V12SP8/20202561

Kamran, M. A., Mannan, M. M. N., & Jeong, M. Y. (2019). Drowsiness, fatigue and poor sleep's causes and detection: A comprehensive study. *IEEE Access: Practical Innovations, Open Solutions*, 7, 167172–167186. doi:10.1109/ACCESS.2019.2951028

Kannapiran, S., & Chakrapani, A. (2017). A novel home automation system using Bluetooth and Arduino. *International Journal of Advances in Computer and Electronics Engineering, 2*(2), 41-44.

Karagiannis, D., & Argyriou, A. (2018). Jamming attack detection in a pair of RF communicating vehicles using unsupervised machine learning. *Vehicular Communications, 13*, 56–63. doi:10.1016/j.vehcom.2018.05.001

Karbab, E., Djenouri, D., Boulkaboul, S., & Bagula, A. (2015, May). Car park management with networked wireless sensors and active RFID. *2015 IEEE International Conference on Electro/Information Technology (EIT)*. 10.1109/EIT.2015.7293372

Kasar, M. V. V. (2019). Smart Bins Concept Implementation in India- Garbage Monitoring System using IOT. *International Journal for Research in Applied Science and Engineering Technology, 7*(6), 1939–1942. doi:10.22214/ijraset.2019.6326

Kashan Ali Shah, S., & Mahmood, W. (2020). Smart Home Automation Using IOT and its Low Cost Implementation. *International Journal of Engineering and Manufacturing, 10*(5), 28–36. doi:10.5815/ijem.2020.05.03

Kaur, M., Kaur, S., & Singh, G. (2012). Vehicular ad hoc networks. *J. Glob. Res. Comput. Sci., 3*(3), 61–64. https://www.researchgate.net/publication/279973104_

Kavitha, M. N., Saranya, S. S., Adithyan, K. D., Soundharapandi, R., & Vignesh, A. S. (2021, November). A novel approach for driver drowsiness detection using deep learning. In. AIP Conference Proceedings: Vol. 2387. *No. 1* (p. 140027). AIP Publishing LLC. doi:10.1063/5.0068784

Khan, U. A., & Lee, S. S. (2019). Multi-layer problems and solutions in VANETs: A review. *Electronics (Basel), 8*(2), 204. doi:10.3390/electronics8020204

Khanum, A., & Shivakumar, R. (2019). An enhanced security alert system for smart home using IOT. *Indonesian Journal of Electrical Engineering and Computer Science, 13*(1), 27. doi:10.11591/ijeecs.v13.i1.pp27-34

Khelifi, A., Abu Talib, M., Nouichi, D., & Eltawil, M. S. (2019). Toward an Efficient Deployment of Open Source Software in the Internet of Vehicles Field. *Arabian Journal for Science and Engineering, 44*(11), 8939–8961. doi:10.100713369-019-03870-2

Khoa, T. A., Man, M. M., Nguyen, T. Y., Nguyen, V., & Nam, N. H. (2019). Smart Agriculture Using IoT Multi-Sensors: A Novel Watering Management System. *Journal of Sensor and Actuator Networks, 8*(3), 45. doi:10.3390/jsan8030045

Kim, S., Kim, W., Noh, Y., & Park, F. C. (2017). Transfer learning for automated optical inspection. *2017 International Joint Conference on Neural Networks (IJCNN)*, 2517–2524. 10.1109/IJCNN.2017.7966162

Kornblith, S., Shlens, J., & Le, Q. V. (2019). Do better ImageNet models transfer better? *2019 IEEE/CVF Conference on Computer Vision and Pattern Recognition (CVPR)*, 2656–2666. 10.1109/CVPR.2019.00277

Kristen, E., Kloibhofer, R., Díaz, V. H., & Castillejo, P. (2021). Security Assessment of Agriculture IoT (AIoT) Applications. *Applied Sciences (Basel, Switzerland)*, *11*(13), 5841. doi:10.3390/app11135841

Kumar, A., Kumar, A., Singh, A. K., & Choudhary, A. K. (2021). IoT Based Energy Efficient Agriculture Field Monitoring and Smart Irrigation System using NodeMCU. *Journal of Mobile Multimedia*. doi:10.13052/jmm1550-4646.171318

Kumar, S., & Lee, S. R. (2014, June). Android based smart home system with control via Bluetooth and internet connectivity. In *The 18th IEEE International Symposium on Consumer Electronics (ISCE 2014)* (pp. 1-2). IEEE.

Kumar, P. P., Prasanth, P. P., Hemalatha, P., & Kulakarni, K. J. (2021). A Framework for Fully Automated Home using IoT Reliable Protocol Stack and Smart Gateway. *International Journal of Robotics and Automation Technology*, *7*, 56–62. doi:10.31875/2409-9694.2020.07.7

Kumar, S., & Singh, J. (2020). Internet of Vehicles over VANETs: Smart and secure communication using IoT. *Scalable Computing: Practice and Experience*, *21*(3), 425–440. doi:10.12694cpe.v21i3.1741

Kurniawan, J., Syahra, S. G., Dewa, C. K., & Afiahayati. (2018). Traffic Congestion Detection: Learning from CCTV Monitoring Images using Convolutional Neural Network. *Procedia Computer Science*, *144*, 291–297. doi:10.1016/j.procs.2018.10.530

Laberteaux, K., & Hartenstein, H. (Eds.). (2009). *VANET: Vehicular applications and inter-networking technologies*. John Wiley & Sons.

Lauren, F. (2020). *iPhone manufacturing in China is in limbo amid a coronavirus outbreak.* Available: https://www.cnbc.com/2020/02/10/coronavirus-leaves-status-of-apple-manufacturing-in-china-uncertain.html

Lazebnik, S. (2015). Deep convolutional neural networks for hyperspectral image classification. *Journal of Sensors*, *2015*, 258619.

Lee, E.-K., Gerla, M., Pau, G., Lee, U., & Lim, J.-H. (2016, September). Internet of Vehicles: From intelligent grid to autonomous cars and vehicular fogs. *International Journal of Distributed Sensor Networks*, *12*(9). Advance online publication. doi:10.1177/1550147716665500

Li, A., Bodanese, E., Poslad, S., Hou, T., Wu, K., & Luo, F. (2022). A Trajectory-based Gesture Recognition in Smart Homes based on the Ultra-Wideband Communication System. *IEEE Internet of Things Journal*, *1*, 1. Advance online publication. doi:10.1109/JIOT.2022.3185084

Liang, Li, Zhang, Wang, & Bie. (2015). Vehicular Ad Hoc Networks: Architectures, Research Issues, Methodologies, Challenges, and Trends. *International Journal of Distributed Sensor Networks, 5*.

Liang, L., Ye, H., & Li, G. Y. (2019). Toward Intelligent Vehicular Networks: A Machine Learning Framework. *IEEE Internet of Things Journal*, *6*(1), 124–135. doi:10.1109/JIOT.2018.2872122

Lin, H. D., & Ho, D. C. (2007). Detection of pinhole defects on chips and wafers using DCT enhancement in computer vision systems. *International Journal of Advanced Manufacturing Technology*, *34*(5-6), 567–583. doi:10.100700170-006-0614-3

Lin, J., Long, W., Zhang, A., & Chai, Y. (2020). Blockchain and IoT-based architecture design for intellectual property protection. *International Journal of Crowd Science*, *4*(3), 283–293. doi:10.1108/IJCS-03-2020-0007

Lin, K., Li, C., Pace, P., & Fortino, G. (2020). Multi-level cluster-based satellite-terrestrial integrated communication in Internet of vehicles. *Computer Communications*, *149*, 44–50. doi:10.1016/j.comcom.2019.10.009

Li, Y., & Wu, H. (2012). A Clustering Method Based on K-Means Algorithm. *Physics Procedia*, *25*, 1104–1109. doi:10.1016/j.phpro.2012.03.206

Li, Z., Liu, F., Yang, W., Peng, S., & Zhou, J. (2021). A survey of convolutional neural networks: Analysis, applications, and prospects. *IEEE Transactions on Neural Networks and Learning Systems*, 1–21. doi:10.1109/TNNLS.2021.3084827 PMID:34111009

Lubna, M., Mufti, N., & Shah, S. A. A. (2021). Automatic Number Plate Recognition:A Detailed Survey of Relevant Algorithms. *Sensors (Basel)*, *21*(9), 3028. doi:10.339021093028 PMID:33925845

Lund, S., Madgavkar, A., Manyika, J., Smit, S., Ellingrud, K., & Robinson, O. (2020). *The future of work after COVID-19*. Retrieved March 12, 2022, from https://www.mckinsey.com/featured-insights/future-of-work/the-future-of-work-after-covid-19

Luo, H., Xiong, C., Fang, W., Love, P. E., Zhang, B., & Ouyang, X. (2018). Convolutional neural networks: Computer vision-based workforce activity assessment in construction. *Automation in Construction*, *94*, 282–289. doi:10.1016/j.autcon.2018.06.007

Machine Learning Prediction Analysis using IoT for Smart Farming. (2020). *International Journal of Emerging Trends in Engineering Research*, *8*(9), 6482–6487. doi:10.30534/ijeter/2020/250892020

Magán, E., Sesmero, M. P., Alonso-Weber, J. M., & Sanchis, A. (2022). Driver Drowsiness Detection by Applying Deep Learning Techniques to Sequences of Images. *Applied Sciences (Basel, Switzerland)*, *12*(3), 1145. doi:10.3390/app12031145

Maheshwaran, K. (2018, April 24). Smart Garbage Monitoring System using IOT. *IJERT*. https://www.ijert.org/smart-garbage-monitoring-system-using-iot

Mahindrakar, S., & Biradar, R. K. (2017a). Internet of Things: Smart Home Automation System using Raspberry Pi. *International Journal of Scientific Research*, *6*(1), 901–905. doi:10.21275/ART20164204

Mahmoodi, F., Blutinger, E., Echazú, L., & Nocetti, D. (2021). *COVID-19 and the health care supply chain: impacts and lessons learned.* Retrieved April 22, from https://www.supplychainquarterly. com/articles/4417-covid-19-and-the-health-care-supply-chain-impacts-and-lessons-learned

Maier, M. W., Emery, D., & Hilliard, R. (2001). Software architecture: Introducing IEEE standard 1471. *Computer, 34*(4), 107–109. doi:10.1109/2.917550

Maier, M. W., Emery, D., & Hilliard, R. (2002, August). 5.4. 3 ANSI/IEEE 1471 and Systems Engineering. In *INCOSE International Symposium* (Vol. 12, No. 1, pp. 798-805). 10.1002/j.2334-5837.2002.tb02541.x

Malik, I., & Tarar, S. (2021). Cloud-Based Smart City Using Internet of Things. *Integration and Implementation of the Internet of Things Through Cloud Computing*, 133–154. doi:10.4018/978-1-7998-6981-8.ch007

Mamun, M.A., Hannan, M.A., & Hussain, A. (2014). *A novel prototype and simulation model for real time solid wastebin monitoring system.* Academic Press.

Manchanda, S., & Sharma, S. (2017). Extraction and Enhancement of Moving Objects in a Video. *Advances in Computer and Computational Sciences*, 763–771. doi:10.1007/978-981-10-3770-2_72

Mancini, A., Frontoni, E., & Zingaretti, P. (2018). Improving Variable Rate Treatments by Integrating Aerial and Ground Remotely Sensed Data. *2018 International Conference on Unmanned Aircraft Systems (ICUAS)*. 10.1109/ICUAS.2018.8453327

Mani, A., & Charan, R. (2018). A survey on IoT based system for smart home automation and theft control. *International Journal of Modern Trends in Engineering & Research, 5*(2), 70–74. doi:10.21884/IJMTER.2018.5038.VVILQ

Mardi, Z., Ashtiani, S. N., & Mikaili, M. (2011). EEG-based somnolence detection for safe driving exploitation chaotic options and applied math tests. *Journal of Medical Signals and Sensors, 1*(2), 130–137. doi:10.4103/2228-7477.95297 PMID:22606668

Masmoudi, I., Wali, A., Jamoussi, A., & Alimi, A. M. (2014). Vision based System for Vacant Parking Lot Detection: VPLD. *Proceedings of the 9th International Conference on Computer Vision Theory and Applications*, 526–533. 10.5220/0004730605260533

Mathijssen, A., & Pretorius, A. J. (2007). Verified Design of an Automated Parking Garage. *Formal Methods: Applications and Technology*, 165–180. doi:10.1007/978-3-540-70952-7_11

Mayub, A., Fahmizal, F., Shidiq, M., Oktiawati, U. Y., & Rosyid, N. R. (2019). Implementation smart home using internet of things. *TELKOMNIKA, 17*(6), 3126. doi:10.12928/telkomnika. v17i6.11722

Mchergui, A., Moulahi, T., Ben Othman, M. T., & Nasri, S. (2020). Enhancing VANETs broadcasting performance with mobility prediction for smart road. *Wireless Personal Communications, 112*(3), 1629–1641. doi:10.100711277-020-07119-2

Medina, R., Gayubo, F., González, L. M., Olmedo, D., Gómez-García-Bermejo, J., Zalama, E., & Perán, J. (2011). Automated visual classification of frequent defects in flat steel coils. *International Journal of Advanced Manufacturing Technology*, *57*(9-12), 1087–1097. doi:10.100700170-011-3352-0

Mekala, M. S., & Viswanathan, P. (2017). A Survey: Smart agriculture IoT with cloud computing. *2017 International Conference on Microelectronic Devices, Circuits and Systems (ICMDCS)*. 10.1109/ICMDCS.2017.8211551

Mera, C., Orozco-Alzate, M., Branch, J., & Mery, D. (2016). Automatic visual inspection: An approach with multi-instance learning. *Computers in Industry*, *83*, 46–54. doi:10.1016/j.compind.2016.09.002

Mogili, U. R., & Deepak, B. B. V. L. (2018). Review on Application of Drone Systems in Precision Agriculture. *Procedia Computer Science*, *133*, 502–509. doi:10.1016/j.procs.2018.07.063

Mohamad, Abidin, Elias, & Zainol. (2017). *Optimizing Congestion Control for Non-Safety Messages in VANETs Using Taguchi Method*. Academic Press.

Mohan, D., Tsimhoni, O., Sivak, M., & Flannagan, M. J. (2009). *Road safety in India: Challenges and Opportunities*. University of Michigan, Ann Arbor, Transportation Research Institute. https://deepblue.lib.umich.edu/handle/2027.42/61504

Mohanty, S. P. (2015). IEEE ISCE. *IEEE Consumer Electronics Magazine*, *4*(1), 111. doi:10.1109/MCE.2014.2361003

Mormont, R., Geurts, P., & Marée, R. (2018). Comparison of deep transfer learning strategies for digital pathology. *2018 IEEE/CVF Conference on Computer Vision and Pattern Recognition Workshops (CVPRW)*, 2343–234309. 10.1109/CVPRW.2018.00303

Mustafa, M. R. (2017). Smart Bin: Internet-of-Things Garbage Monitoring System. *MATEC Web of Conferences*. Https://Doi.Org/10.1051/Matecconf/201714001030

Na, S., Xumin, L., & Yong, G. (2010, April). Research on k-means Clustering Algorithm: An Improved k-means Clustering Algorithm. *2010 Third International Symposium on Intelligent Information Technology and Security Informatics*. 10.1109/IITSI.2010.74

Naser, M. A. U., Jasim, E. T., & Al-Mashhadi, H. M. (2020). QR code based two-factor authentication to verify paper-based documents. *TELKOMNIKA*, *18*(4), 1834. doi:10.12928/telkomnika.v18i4.14339

National Informatics Centre(NIC), Government of India. (n.d.). *Technology based startups played a crucial role in converting India from importer to second largest manufacturer of PPEs*. Retrieved 31 December, 2020, from https://dst.gov.in/technology-based-startups-played-crucial-role-converting-india-importer-second-largest-manufacturer

Nguyen, D. B., Dow, C. R., & Hwang, S. F. (2018). An Efficient Traffic Congestion Monitoring System on Internet of Vehicles. *Wireless Communications and Mobile Computing*, *2018*, 1–17. doi:10.1155/2018/9136813

Noori, S.M., & Mikaeili, M. (2016). Driving temporary status detection victimization blending of electroencephalography, electrooculography, and driving nature signals. *J Master's Degree Signals Sens, 6*, 39-46. doi:10.4103/2228-7477.175868

Pala, Z., & Inanc, N. (2007, September). Smart Parking Applications Using RFID Technology. *2007 1st Annual RFID Eurasia*. doi:10.1109/RFIDEURASIA.2007.4368108

Palomino, W., Morales, G., Huaman, S., & Telles, J. (2018). PETEFA: Geographic Information System for Precision Agriculture. *2018 IEEE XXV International Conference on Electronics, Electrical Engineering and Computing (INTERCON)*. 10.1109/INTERCON.2018.8526414

Pandya, B., Mehta, M., Jain, N., & Kadam, S. (2016). Android Based Home Automation System Using Bluetooth & Voice Command–Implementation. *International Research Journal of Engineering and Technology*.

Pantelej, E., Gusev, N., Voshchuk, G., & Zhelonkin, A. (2018). Automated Field Monitoring by a Group of Light Aircraft-Type UAVs. *Advances in Intelligent Systems and Computing*, 350–358. doi:10.1007/978-3-030-01821-4_37

Pan, Z., Lie, D., Qiang, L., Shaolan, H., Shilai, Y., Yan-de, L., Yongxu, Y., & Haiyang, P. (2016). Effects of citrus tree-shape and spraying height of small unmanned aerial vehicle on droplet distribution. *International Journal of Agricultural and Biological Engineering, 9*, 45–52.

Park, S., Pan, F., Kang, S., & Yoo, C. D. (2016, November). Driver drowsiness detection system based on feature representation learning using various deep networks. In *Asian Conference on Computer Vision* (pp. 154-164). Springer.

Parraga, A., Doering, D., Atkinson, J. G., Bertani, T., de Oliveira Andrades Filho, C., de Souza, M. R. Q., Ruschel, R., & Susin, A. A. (2018). Wheat Plots Segmentation for Experimental Agricultural Field from Visible and Multispectral UAV Imaging. *Advances in Intelligent Systems and Computing*, 388–399. doi:10.1007/978-3-030-01054-6_28

Pascuzzi, S., Anifantis, A. S., Cimino, V., & Santoro, F. (2018). *Unmanned aerial vehicle used for remote sensing on an Apulian farm in Southern Italy*. Engineering for Rural Development. doi:10.22616/ERDev2018.17.N175

Patel, H. K., Gaur, V., Kumar, S., Singh, A. K., & Mittal, S. (2021). Review on Next Step Home Automation Using Wi-Fi Module. In *Advances in Smart Communication and Imaging Systems* (pp. 301–307). Springer.

Patel, N. J., & Jhaveri, R. H. (2015). Trust based approaches for secure routing in VANET: A Survey. *Procedia Computer Science, 45*, 592–601. doi:10.1016/j.procs.2015.03.112

Patil, D. (2015). Tumor Size Processing using Smart Phone. *International Journal on Recent and Innovation Trends in Computing and Communication, 3*(2), 785–788. doi:10.17762/ijritcc2321-8169.150275

Patil, D. S. N., & Jadhav, M. B. (2019). Smart Agriculture Monitoring System Using IOT. *IJARCCE, 8*(4), 116–120. doi:10.17148/IJARCCE.2019.8419

Perez, L., & Wang, J. (2017). *The effectiveness of data augmentation in image classification using deep learning.* ArXiv, abs/1712.04621.

Piyare, R. (2013). Internet of things: ubiquitous home control and monitoring system using android based smart phone. *International Journal of Internet of Things, 2*(1), 5-11.

Piyare, R., & Tazil, M. (2011, June). Bluetooth based home automation system using cell phone. In *2011 IEEE 15th International Symposium on Consumer Electronics (ISCE)* (pp. 192-195). IEEE.

Plotegher, L., Corridori, C., Dolci, F., Andreatta, C., Benini, S., & Devilli, M. (2016). *The GI dataset for glass inspection 1st release (August 2016). Technical report.* Deltamax Automazione Srl.

Pouyanfar, S., Sadiq, S., Yan, Y., Tian, H., Tao, Y., Reyes, M. P., Shyu, M. L., Chen, S. C., & Iyengar, S. S. (2018). A Survey on Deep Learning: Algorithms, Techniques, and Applications. *ACM Computing Surveys, 51*, 1–36.

Pramanik, A., Nagar, V., Dwivedi, S., & Choudhury, B. (2016, March). GSM based Smart home and digital notice board. In *2016 International Conference on Computational Techniques in Information and Communication Technologies (ICCTICT)* (pp. 41-46). IEEE.

Prathibha, S. R., Hongal, A., & Jyothi, M. P. (2017). IOT Based Monitoring System in Smart Agriculture. *2017 International Conference on Recent Advances in Electronics and Communication Technology (ICRAECT).* 10.1109/ICRAECT.2017.52

Praveen, V., Arunprasad, T., & Gomathi, R. N. D. P. (2018). A Survey on Trash Monitoring System using Internet of Things. *International Journal of Trend in Scientific Research and Development, 2*(3), 601–605. doi:10.31142/ijtsrd11020

Preet, E., Dhindsa, S. K., & Khanna, R. (2017). Encryption Based Authentication Schemes in Vehicular Ad Hoc Networks. *Conference: All India Seminar on Recent Innovations in Wireless Communication & Networking At: Baba Banda Singh Bahadur Engineering College Fateh garh Sahib,* 15–21.

Radoglou-Grammatikis, P., Sarigiannidis, P., Lagkas, T., & Moscholios, I. (2020). A compilation of UAV applications for precision agriculture. *Computer Networks, 172,* 107148. doi:10.1016/j. comnet.2020.107148

Rafi, F. A. (2018). Implementation of smart home automation using raspberry pi. *International Journal of Recent Trends in Engineering and Research,* 122–125. Advance online publication. doi:10.23883/IJRTER.CONF.02180328.019.HDV3S

Rahayu, Y., & Mustapa, F. N. (2013). A Secure Parking Reservation System Using GSM Technology. *International Journal of Computer and Communication Engineering,* 518–520. Advance online publication. doi:10.7763/IJCCE.2013.V2.239

Raheem, A. K. K. A. (2017). Bluetooth based smart home automation system using arduino UNO microcontroller. *Al-Mansour Journal, 27,* 139–156.

Raju, G. S., Kumar, N. S., & Nikkat, S. (2020, December). Technology based startups pivoting for sustainability: Case study of startups. *IOP Conference Series. Materials Science and Engineering*, *981*(2), 022083. doi:10.1088/1757-899X/981/2/022083

Ramesh, M., Vijay Kumar, G., Suresh Babu, B., Boopathi, R., Sreekanth, C., Muthukumar, P., & Padma Suresh, L. (2021). Exploration or Multipurpose Electric Vehicle for Agriculture Using IOT. *Tobacco Regulatory Science*, *7*(5), 3844–3852. doi:10.18001/TRS.7.5.1.157

Ramlee, R. A., Othman, M. A., Leong, M. H., Ismail, M. M., & Ranjit, S. S. S. (2013, March). Smart home system using android application. In *2013 International Conference of Information and Communication Technology (ICoICT)* (pp. 277-280). IEEE.

Ransing, R. S., & Rajput, M. (2015, January). Smart home for elderly care, based on Wireless Sensor Network. In *2015 International Conference on Nascent Technologies in the Engineering Field (ICNTE)* (pp. 1-5). IEEE.

Rathod, N. (2020). Smart Farming: IOT Based Smart Sensor Agriculture Stick for Live Temperature and Humidity Monitoring. *International Journal of Engine Research*, *V9*(07). Advance online publication. doi:10.17577/IJERTV9IS070175

Raya. (2007). *On data-centric trust establishment in ephemeral ad hoc networks* (Technical Report). LCA-REPORT-2007-003.

Ren, R., Hung, T., & Tan, K. (2018). A generic deep learning-based approach for automated surface inspection. *IEEE Transactions on Cybernetics*, *48*(3), 929–940. doi:10.1109/TCYB.2017.2668395 PMID:28252414

Risteska Stojkoska, B. L., & Trivodaliev, K. V. (2017). A review of Internet of Things for smart home: Challenges and solutions. *Journal of Cleaner Production*, *140*, 1454–1464. doi:10.1016/j.jclepro.2016.10.006

Romeo, M. S. S. (2019). Intrusion Detection System (IDS) in Internet of Things (IoT) Devices for Smart Home. *International Journal of Psychosocial Rehabilitation*, *23*(4), 1217–1227. doi:10.37200/IJPR/V23I4/PR190448

Ryu, S., Kim, K., Kim, J. Y., Cho, I. K., Kim, H., Ahn, J., Choi, J., & Ahn, S. (2022). Design and Analysis of a Magnetic Field Communication System Using a Giant Magneto-Impedance Sensor. *IEEE Access: Practical Innovations, Open Solutions*, *10*, 56961–56973. doi:10.1109/ACCESS.2022.3171581

S., K. (2018). IoT in Agriculture: Smart Farming. *International Journal of Scientific Research in Computer Science, Engineering and Information Technology*, 181–184. doi:10.32628/CSEIT183856

S., L., & B., H. (2018). Design and Implementation of IOT based Smart Security and Monitoring for Connected Smart Farming. *International Journal of Computer Applications, 179*(11), 1–4. doi:10.5120/ijca2018914779

Salaan, C. J., Tadakuma, K., Okada, Y., Sakai, Y., Ohno, K., & Tadokoro, S. (2019). Development and Experimental Validation of Aerial Vehicle with Passive Rotating Shell on Each Rotor. *IEEE Robotics and Automation Letters, 4*(3), 2568-2575. doi:10.1109/LRA.2019.2894903

Samann, F. E. F. (2017, August 30). The Design and Implementation of Smart Trash Bin. *Academic Journal of Nawroz University.* Https://Doi.Org/10.25007/Ajnu.V6n3a103. https://journals.nawroz.edu.krd/index.php/ajnu/article/view/103

Sangeetha, S. B. (2015, March). Intelligent interface based speech recognition for home automation using android application. In *2015 International Conference on Innovations in Information, Embedded and Communication Systems (ICIIECS)* (pp. 1-11). IEEE.

Sanger, J. B., Sitanayah, L., & Ahmad, I. (2021). A Sensor-based Trash Gas Detection System. *2021 IEEE 11th Annual Computing and Communication Workshop and Conference (CCWC).* 10.1109/CCWC51732.2021.9376147

Schilirò, D. (2020). *Towards digital globalization and the covid-19 challenge.* Academic Press.

Schroth, C., Kosch, T., Strassberger, M., & Bechler, M. (2012). *Automotive internetworking.* John Wiley & Sons.

Senouci, O., Aliouat, Z., & Harous, S. (2018). A review of routing protocols in internet of vehicles and their challenges. *Sensor Review.*

Sen, S., Chakrabarty, S., Toshniwal, R., & Bhaumik, A. (2015). Design of an intelligent voice controlled home automation system. *International Journal of Computers and Applications, 121*(15).

Serpen, G., & Debnath, J. (2019). Design and performance evaluation of a parking management system for automated, multi-story and robotic parking structure. *International Journal of Intelligent Computing and Cybernetics, 12*(4), 444–465. doi:10.1108/IJICC-02-2019-0017

Serrão, C., & Garrido, N. (2019). A Low-Cost Smart Parking Solution for Smart Cities Based on Open Software and Hardware. *Lecture Notes of the Institute for Computer Sciences, Social Informatics and Telecommunications Engineering,* 15–25. doi:10.1007/978-3-030-14757-0_2

Sharma, S., Ghanshala, K. K., & Mohan, S. (2018, November). A Security System Using Deep Learning Approach for Internet of Vehicles (IoV). *2018 9th IEEE Annual Ubiquitous Computing, Electronics & Mobile Communication Conference (UEMCON).* doi:10.1109/UEMCON.2018.8796664

Shinde, S. S., Yadahalli, R. M., & Shabadkar, R. (2021). Cloud and IoT-Based Vehicular Ad Hoc Networks (VANET). *Cloud and IoT-Based Vehicular Ad Hoc Networks,* 67-82.

Shinde, S. S., & Patil, S. P. (2010, July–December). Various Issues in Vehicular Ad Hoc Networks: A Survey. *Int. J. Comput. Sci. Commun., 1*(2), 399–403.

Shinde, S. S., Yadahalli, R. M., & Tamboli, A. S. (2012, April–June). Vehicular Ad Hoc Network Localization Techniques: A Review. *Int. J. Electron. Commun. Technol., 3*(2), 82–86.

Shorten, C., & Khoshgoftaar, T. (2019). A survey on image data augmentation for deep learning. *Journal of Big Data*, 6(1), 1–48. doi:10.118640537-019-0197-0

Shukla, A. (2020). *Coronavirus fears send the aerospace industry into a tailspin*. Available: https://www.rediff.com/business/report/covid-19-fears-send-aerospace-industry-into-tailspin/20200322.htm

Sidhu, N., Pons-Buttazzo, A., Muñoz, A., & Terroso-Saenz, F. (2021). A Collaborative Application for Assisting the Management of Household Plastic Waste through Smart-Bins: A Case of Study in the Philippines. *Sensors (Basel)*, 21(13), 4534. doi:10.339021134534 PMID:34282802

Singh, A. K., Tamta, P., & Singh, G. (2019). Smart Parking System using IoT. *International Journal of Engineering and Advanced Technology*, 9(1), 6091–6095. doi:10.35940/ijeat.A1963.109119

Singh, M. K., Singh, R., Singh, N., & Yadav, C. S. (2022). Technologies Assisting the Paradigm Shift from 5G to 6G. In *AI and Blockchain Technology in 6G Wireless Network* (pp. 1–24). Springer. doi:10.1007/978-981-19-2868-0_1

Singh, P., Singh, N., & Deka, G. C. (2021). Prospects of machine learning with blockchain in healthcare and agriculture. In *Multidisciplinary functions of Blockchain technology in AI and IoT applications* (pp. 178–208). IGI Global. doi:10.4018/978-1-7998-5876-8.ch009

Singh, P., Singh, N., Ramalakshmi, P., & Saxena, A. (2022). Artificial Intelligence for Smart Data Storage in Cloud-Based IoT. In *Transforming Management with AI, Big-Data, and IoT* (pp. 1–15). Springer. doi:10.1007/978-3-030-86749-2_1

Singh, P., Singh, N., Singh, K. K., & Singh, A. (2021). Diagnosing of disease using machine learning. In *Machine Learning and the Internet of Medical Things in Healthcare* (pp. 89–111). Academic Press. doi:10.1016/B978-0-12-821229-5.00003-3

Smith, M. L., Smith, L. N., & Hansen, M. F. (2021). The quiet revolution in machine vision - a state-of-the-art survey paper, including historical review, perspectives, and future directions. *Computers in Industry*, 130, 103472. doi:10.1016/j.compind.2021.103472

Soleymani, A., Abdullah, A. H., Hassan, W. H., Anisi, M. H., Goudarzi, S., Rezazadch Baee, M. A., & Mandala, S. (2015). Trust Management in Vehicular Ad-hoc Network. *EURASIP Journal on Wireless Communications and Networking*, 2015(1), 146. doi:10.118613638-015-0353-y

Song, Z., Yuan, Z., & Liu, T. (2019). Residual squeezeand-excitation network for battery cell surface inspection. *2019 16th International Conference on Machine Vision Applications (MVA)*, 1–5.

Souza, I. R., Escarpinati, M. C., & Abdala, D. D. (2018). A curve completion algorithm for agricultural planning. *Proceedings of the 33rd Annual ACM Symposium on Applied Computing*. 10.1145/3167132.3167158

Srisabarimani, K., Arthi, R., Soundarya, S., Akshaya, A., Mahitha, M., & Sudheer, S. (2022a). Smart Segregation Bins for Cities Using Internet of Things (IoT). *Lecture Notes in Electrical Engineering*, 717–730. doi:10.1007/978-981-16-9488-2_68

Srivastava, A., Prakash, A., & Tripathi, R. (2020). Location based routing protocols in VANET: Issues and existing solutions. *Vehicular Communications*, *23*, 100231. doi:10.1016/j.vehcom.2020.100231

Stentoft, J., & Rajkumar, C. (2020). The relevance of Industry 4.0 and its relationship with moving to manufacture out, back, and staying at home. *International Journal of Production Research*, *58*(10), 2953–2973. doi:10.1080/00207543.2019.1660823

Suddul, G. (2018). An Energy Efficient and Low Cost Smart Recycling Bin. *International Journal of Computer Applications*. https://www.ijcaonline.org/archives/volume180/number29/29177-2018916698

Sushanth, G., & Sujatha, S. (2018). IOT Based Smart Agriculture System. *2018 International Conference on Wireless Communications, Signal Processing and Networking (WiSPNET)*. 10.1109/WiSPNET.2018.8538702

Szegedy, C., Ioffe, S., Vanhoucke, V., & Alemi, A. A. (2017). Inception-V4, inception-ResNet and the impact of residual connections on learning. In *Proceedings of the Thirty-First AAAI Conference on Artificial Intelligence* (pp. 4278–4284). AAAI Press. 10.1609/aaai.v31i1.11231

Tabernik, D., Šela, S., Skvarč, J., & Skočaj, D. (2019). Deep-learning-based computer vision system for surface-defect detection. In D. Tzovaras, D. Giakoumis, M. Vincze, & A. Argyros (Eds.), *Computer Vision Systems* (pp. 490–500). Springer International Publishing. doi:10.1007/978-3-030-34995-0_44

Takahashi, R., Matsubara, T., & Uehara, K. (2020). Data augmentation using random image cropping and patching for deep CNNs. *IEEE Transactions on Circuits and Systems for Video Technology*, *30*(9), 2917–2931. doi:10.1109/TCSVT.2019.2935128

Takase, T., Karakida, R., & Asoh, H. (2021). Self-paced data augmentation for training neural networks. *Neurocomputing*, *442*, 296–306. doi:10.1016/j.neucom.2021.02.080

Talib, Hussin, & Hassan. (2017). Converging VANET with Vehicular Cloud Networks to reduce the Traffic Congestions: A review. *International Journal of Applied Engineering Research*, *12*(21), 10646-10654.

Tanuja, K., Sushma, T.M., Bharati, M., & Arun, K.H. (2015). A Survey on VANET Technologies. *Int. J. Comput. Appl., 121*(18), 1–9.

Tapakire. (2019). IoT based Smart Agriculture using Thingspeak. *International Journal of Engineering Research And, 8*(12). doi:10.17577/IJERTV8IS120185

Terry, M. (2018). *Recent drug scandals in China spotlight potential global supply chain issues*. Available: https://www.biospace.com/article/recent-drug-scandals-in-china-spotlight-potential-global-supply-chain-issues/

Teymourzadeh, R., Ahmed, S. A., Chan, K. W., & Hoong, M. V. (2013, December). Smart GSM based home automation system. In *2013 IEEE Conference on Systems, Process & Control (ICSPC)* (pp. 306-309). IEEE.

The World Bank. (n.d.). *Food Security and Covid.* Retrieved January 20th, 2022, from https://www.worldbank.org/en/topic/agriculture/brief/food-security-update

Turner, D., Lucieer, A., & Watson, C.S. (2011). *Development of an Unmanned Aerial Vehicle (UAV) for hyper-resolution vineyard mapping based on visible, multispectral and thermal imagery.* Academic Press.

Tzounis, A., Katsoulas, N., Bartzanas, T., & Kittas, C. (2017). Internet of Things in agriculture, recent advances and future challenges. *Biosystems Engineering*, *164*, 31–48. doi:10.1016/j.biosystemseng.2017.09.007

Ullah, M. A., & Celik, A. R. (2016). An effective approach to build smart building based on Internet of Things (IoT). *Journal of Basic and Applied Scientific Research*, *6*, 56–62.

Verma, A., Singh, P., & Singh, N. (2021). Study of blockchain-based 6G wireless network integration and consensus mechanism. *International Journal of Wireless and Mobile Computing*, *21*(3), 255–264. doi:10.1504/IJWMC.2021.120906

Villa-Henriksen, A., Edwards, G. T., Pesonen, L. A., Green, O., & Sørensen, C. A. G. (2020). Internet of Things in arable farming: Implementation, applications, challenges and potential. *Biosystems Engineering*, *191*, 60–84. doi:10.1016/j.biosystemseng.2019.12.013

Voulodimos, A., Doulamis, N., Doulamis, A., & Protopapadakis, E. (2018). Deep Learning for Computer Vision: A Brief Review. *Computational Intelligence and Neuroscience*, *2018*, 13. doi:10.1155/2018/7068349 PMID:29487619

Wamba, S. F., Dubey, R., Gunasekaran, A., & Akter, S. (2020). The performance effects of big data analytics and supply chain ambidexterity: The moderating effect of environmental dynamism. *International Journal of Production Economics*, *222*, 107498. doi:10.1016/j.ijpe.2019.09.019

Wang, T., Cao, Z., Wang, S., Wang, J., Qi, L., Liu, A., Xie, M., & Li, X. (2020). Privacy-Enhanced Data Collection Based on Deep Learning for Internet of Vehicles. *IEEE Transactions on Industrial Informatics*, *16*(10), 6663–6672. doi:10.1109/TII.2019.2962844

Weimer, D., Benggolo, A. Y., & Freitag, M. (2015). Context-aware deep convolutional neural networks for industrial inspection. In *Australasian Joint Conference on Artificial Intelligence.* Queensland University of Technology.

Wijaya, A. S. (2017, August 1). *Design a smart waste bin for smart waste management.* IEEE Conference Publication. https://ieeexplore.ieee.org/abstract/document/8068414

Wongcharoen, S., & Senivongse, T. (2016, July). Twitter analysis of road traffic congestion severity estimation. *2016 13th International Joint Conference on Computer Science and Software Engineering (JCSSE).* 10.1109/JCSSE.2016.7748850

World Economic Forum. (2020). *The COVID-19 pandemic has changed education forever. This is how.* Retrieved February 12, 2022, from https://www.weforum.org/agenda/2020/04/coronavirus-education-global-covid19-online-digital-learning/

Xiao, T., Liu, L., Li, K., Qin, W., Yu, N., & Li, Z. (2018). Comparison of transferred deep neural networks in ultrasonic breast masses discrimination. *BioMed Research International, 2018*, 1–9. doi:10.1155/2018/4605191 PMID:30035122

Xu, Z., Elomri, A., Kerbache, L., & El Omri, A. (2020). Impacts of COVID-19 on global supply chains: Facts and perspectives. *IEEE Engineering Management Review, 48*(3), 153–166. doi:10.1109/EMR.2020.3018420

Yan, Z. (2016). Complex Systems Smart Home Security Studies based Big Data Analytics. *International Journal of Smart Home, 10*(6), 41–52. doi:10.14257/ijsh.2016.10.6.05

Yar, H., Imran, A. S., Khan, Z. A., Sajjad, M., & Kastrati, Z. (2021). Towards Smart Home Automation Using IoT-Enabled Edge-Computing Paradigm. *Sensors (Basel), 21*(14), 4932. doi:10.339021144932 PMID:34300671

Yun, J., Li, X., Wei Wang, S. L., & Li, L. (2013). Research and Development of LBS-based Ethnic Cultural Information Resources Smart Recommendation System in Mobile Internet. *Information Technology Journal, 12*(21), 6440–6448. doi:10.3923/itj.2013.6440.6448

Yu, Z., Wu, X., & Gu, X. (2017). Fully convolutional networks for surface defect inspection in industrial environment. In M. Liu, H. Chen, & M. Vincze (Eds.), *Computer Vision Systems* (pp. 417–426). Springer International Publishing. doi:10.1007/978-3-319-68345-4_37

Zamora-Izquierdo, M. A., Santa, J., Martínez, J. A., Martínez, V., & Skarmeta, A. F. (2019). Smart farming IoT platform based on edge and cloud computing. *Biosystems Engineering, 177*, 4–17. doi:10.1016/j.biosystemseng.2018.10.014

Zeadally, S., Chen, Y.-S., Irwin, A., & Hassan, A. (2012). Vehicular ad hoc networks (VANETS): Status, results, and challenges. *Telecommun. Syst., 50*, 217–241. https://link.springer.com/10.1007/s11235-010-9400-5

Zhang, C., Wu, X., Zheng, X., & Yu, S. (2019). Driver drowsiness detection using multi-channel second order blind identifications. *IEEE Access: Practical Innovations, Open Solutions, 7*, 11829–11843. doi:10.1109/ACCESS.2019.2891971

Zhang, J., & Tao, D. (2021). Empowering Things With Intelligence: A Survey of the Progress, Challenges, and Opportunities in Artificial Intelligence of Things. *IEEE Internet of Things Journal, 8*(10), 7789–7817. doi:10.1109/JIOT.2020.3039359

Zhou, T., Ruan, S., & Canu, S. (2019). A review: Deep learning for medical image segmentation using multi-modality fusion. *Array, 3-4*, 100004. doi:10.1016/j.array.2019.100004

Zhou, Z.-H. (2017). A brief introduction to weakly supervised learning. *National Science Review, 5*(1), 44–53. doi:10.1093/nsr/nwx106

Zhu, J., Yuan, Z., & Liu, T. (2019). Welding joints inspection via residual attention network. *2019 16th International Conference on Machine Vision Applications (MVA)*, 1–5. 10.23919/MVA.2019.8758040

About the Contributors

Divya Upadhyay Mishra has worked as a full-time academician for the last 15 years. She works as a Professor and Head of Computer Science & Engineering at ABES Engineering College, Ghaziabad. She has a PhD (Engineering) from Amity University Uttar Pradesh MTech (Information Security) from Guru Gobind Singh Indraprastha University, New Delhi. She also holds B. Tech (Computer Science & Engineering) from Sant Longowal Institute of Engineering & Technology, IKJPTU, Punjab. She has published many research papers in reputed international journals and conferences with Scopus & SCI indexing. IEEE Sensors Council also appreciated her for the paper "Application of Non-Linear Gaussian Regression-Based Adaptive Clock Synchronization Technique for Wireless Sensor Network in Agriculture", IEEE Sensors Journal, Vol. 18, No. 10, May 2018, paper being one of the 25 most downloaded Sensors Journal papers in February 2019. She is an active member of international societies like SM-IEEE and IAENG and an organizing committee for many international conferences, workshops, and seminars to enhance teaching-learning. Her research area includes Networking, Cloud, Security, and IoT.

Shanu Sharma is currently working as an Assistant Professor in Department of Computer Science & Engineering at ABES Engineering College, Ghaziabad (Affiliated to A.P.J Abdul Kalam Technical University, Lucknow). She is having 11.5 years of teaching and research experience. Her research area includes Cognitive computing, Computer Vision, Pattern Recognition and Machine Learning. She has published and presented her work in various National and International Conferences and Journals and currently associated with various reputed International Conferences and journals as Reviewer. She is also serving as a Guest Editor of Special Issue on Intelligent Systems and Application, International Journal of Intelligent Information Technologies (IJIIT), IGI Global and Int. J. of Operations Research and Information Systems (IJORIS), IGI Global. She is a Senior member of IEEE and also an active member of other professional societies like ACM, Soft Computing Research Society and IAENG.

* * *

Ayushi Agarwal is a research scholar at ABES Engineering College, Ghaziabad, India. She completed her M.Tech from AKTU, Lucknow, in the year 2022.

Mahesh B. N. is an M.Com Graduate from Bangalore University, Working in the capacity of Asst Professor in RNS First Grade College, Bangalore from past 10 years, finance is his area of passion. Presented papers at various national conferences.

Vijay Bose is currently working as Assistant Professor in the Department of Business Management, in Vaagdevi College of Engineering at Warangal, Telangana state. He is having Five years of teaching, Four years of Research and Three years of Industry experience. He has published three papers in International Journals, and two Papers in National Journal. His areas of interest are Marketing and Human Resources Management.

Subha K. is presently working as a Teaching Assistant in the Department of Logistics at Alagappa University from 2016 to Present. She has published number of research works in various national and international journals. She also has presented so many papers in the national as well as international Conferences. Her are of interest is Microfinance and Logistics.

Selvaradjou Kandasamy is working as professor in Department of CSE, Pondicherry Technological University, Puducherry. He completed his PH. D from IIT Madras. He is having 25 years of teaching and research experience. He published many research papers in reputed journals and conferences.

Madan Mohan Laddunuri from India worked at several universities in India and abroad. His research papers have appeared in many peer-reviewed international journals. Moreover he published several books related to Social Sciences.

Divya Lanka is pursuing Ph. D in Department of CSE in Pondicherry Engineering College under Pondicherry Central University. He completed M. Tech in 2012 from JNTU Kakinada and B. Tech in 2008 from Andhra University. Her research areas of interest are Internet of Things and Software Defined Networks.

Satheesh Pandian Murugan is presently working as Assistant Professor in the Department of Economics and Centre for Research in Economics, Arumugam Pillai Seethai Ammal College, Tiruppattur, Sivaganga (Dt), Tamil Nadu, India. He is having Nine years of teaching experience along with the M.A., M.Phil., Ph.D., qualifications. He has earned his Ph.D from the Madurai Kamarajar University. He has published more than 10 research papers in his credit both in National and International level. His areas of research interests are Agricultural and Industrial Economics.

Ramakrishna Narasimhaiah is working as an Assistant Professor in the Dept. of Economics, Acharya Institute of Graduate Studies with 13 years of experience in the field of Economics. Prior to this position he was principal in-charge in Siddaganga Shivakumara Swamiji degree College, Bengaluru. He is an active researcher in the filed of Economics, Finance and Insurance. He has been acted as a resource person in the various national and international conferences. He also has published papers in various national and international journals including SSCI and Scopus.

Mauricio Orozco-Alzate received his undergraduate degree in Electronic Engineering, his M.Eng. degree in Industrial Automation and his Dr.Eng. degree in Automatics from Universidad Nacional de Colombia - Sede Manizales, in 2003, 2005 and 2008 respectively. In 2007, he served as a Research Fellow at the Information and Communication Theory Group - Pattern Recognition Laboratory (PRLab) of Delft University of Technology (TU Delft), The Netherlands. Since 2008 he has been with Universidad Nacional de Colombia - Sede Manizales, where he is currently a Full Professor in the Department of Informatics and Computing. His main research interests encompass pattern recognition, digital signal processing and their applications to industrial inspection and environmental data science, the latter particularly for the analysis and classification of seismic, bioacoustic and hydro-meteorological signals.

Poonguzhali S. is currently working as Associate Professor, VISTAS, Chennai and has a rich experience of 15 years in teaching. She completed her Ph.D from Bharathiar University in 2020. Her area of research are Data Mining, Image Processing and Big Data.

Shouvik Sanyal is an Assistant Professor of Marketing in Dhofar University, Sultanate of Oman. HE has over two decades of teaching experience at undergraduate and postgraduate levels in India and Oman. He has authored several research papers in indexed journals in the fields of Marketing and Entrepreneurship. He has also presented papers at conferences in five countries.

Vidhya Shanmugam is presently working as an Assistant Professor in REVA Business School, REVA University- Bangalore. She has 3 years of corporate experience, 5 years of research experience and 5 years of teaching experience. She has completed MBA degree and Ph.D in Human Resource at Bharathiar University, Coimbatore and Data Analytics courses from reputed institutions. Her Current research interests are People Analytics and Artificial Intelligence in Management Fields. She acted as a resource person in different universities and colleges. Worked as an Editor at Shanlax Publications, Published a Book in Lambert Academic Publishing, Germany and several research papers in International Journals out of which six in Scopus Indexed & WoS Journals and received a Best Paper award in National Research Colloquium, 2015.

Pushpa Singh is an Associate Professor in the Department of Computer Science and Information Technology at KIET Group of institutions, Delhi-NCR, Ghaziabad, India. She has more than 18 years of experience teaching B.Tech and MCA students. Dr. Pushpa has acquired MCA, M.Tech (CSE), and Ph.D. (CSE) in Wireless Networks from AKTU Lucknow. Her current areas of research include performance evaluation of heterogeneous wireless networks, machine learning, and cryptography. She has published 45 papers in reputed international journals and conference proceedings. She has published four books and contributed six book chapters in international publication. She is one of active reviewers of IGI, Springer, and other conference proceedings and journals. She has been invited to serve on various technical program committee.

Dhanabalan Thangam presently working as Assistant Professor in Commerce and Management at Acharya Institute of Graduate Studies, Banglore, India. Earlier he was worked as Post - Doctoral researcher in Konkuk School of Business, Konkuk University, Seoul, Korea South. He received his Ph.D. degree in Management from Alagappa University, Tamilnadu, India. His current research interests are marketing, small business management and Industry 4.0 Technologies and its application in Business and Management.

D. S. Vijayan (Academician, Researcher, Reviewer, Editor, Innovator and Consultant) completed Graduation in Civil Engineering from Bharath University in the year 2007, Post-Graduation in Structural Engineering from B.S Abdur Rahman Crescent Engineering College Affiliated to Anna University in the year 2010 and has received Doctorate degree from Bharath Institute of Higher Education and Research in June 2017 in the title of "Behavior Of Pre-Stressed Concrete Beams Strengthened With Fibre Reinforced Polymer Laminates" with high recommendation and completed my doctorate in pre-stressed concrete beams with FRP Laminates and has major interest in design and experimental works. Published 55 International Research papers and 12 National Research Paper. He has 6 Patent to his credit in the area of Juliflora plant Ash replacement with cement in the field of civil engineering and attended

23 National and International workshop. Attended more than 100 online webinars and complete 40 online courses in courser, etc. Reviewed 53 research article from leading journals like Elsevier journals, Scopus etc. Have 2 years of industrial and 11 years of teaching experience. Has a life member in ICI (Indian Concrete institute), ISTE (Indian Society for Technical Education), InSc Life member, Institution of Engineers (India) [IEI] life member from 2019 and ASCE (American Society for Civil Engineering). Certified Registered Engineer Grade I from Greater Chennai Corporation from 2019 to 2024 and Certified Registered Structural Engineer Grade I from CMDA (Chennai Metropolitan Development Authority) from 2019 to 2024. Certified Chartered Engineer from Institution of Engineers (India) [IEI]. Currently working as Associate Professor in the department of Civil Engineering on Aarupadi Veedu institute of Technology and guiding 2-Part time Research Scholar Currently working in the field of sensor Structural Heath Monitoring, Geopolymer Stabilized Expansive Soils, Bottom Ash, Pre-Stressed Concrete, etc.

Eduardo-José Villegas-Jaramillo received his undergraduate degree in System Engineering from Universidad Autónoma de Manizales, in 1989 and his M.Eng. degree in System Engineering and Computing from Universidad de los Andes in 1991. Since 2016 he has been pursuing his doctoral degree in Engineering – Industrial and Organizations at Universidad Nacional de Colombia - Sede Manizales, where he is also an Associate Professor in the Department of Informatics and Computing. His main research interests encompass pattern recognition, machine learning, image processing and their applications to industrial inspection.

Index

Index

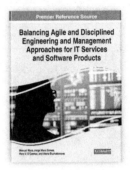

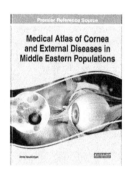
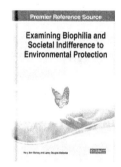
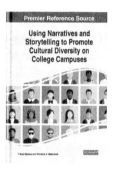

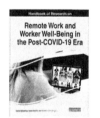

Printed in the United States
by Baker & Taylor Publisher Services